Notebo... ...blems

New technologies and development

Science and technology as factors of change: impact of recent and foreseeable scientific and technological progress on the evolution of societies, especially in the developing countries

Edited by Ann Johnston
and Albert Sasson

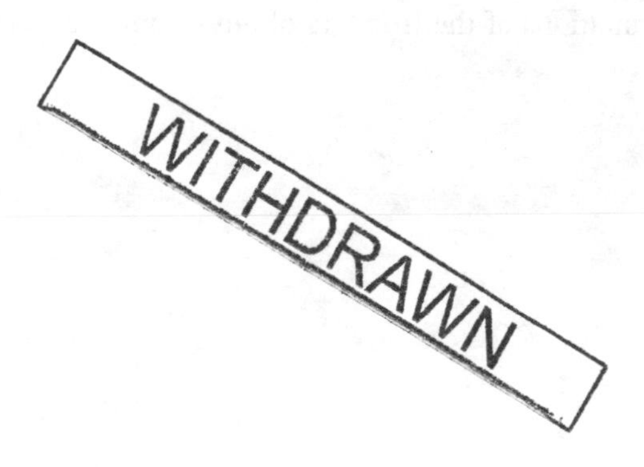

Unesco

The designations employed and the presentation of the material in this publication do not imply the expression of any opinion whatsoever on the part of the Unesco Secretariat concerning the legal status of any country or territory, or of its authorities, or concerning the delimitations of the frontiers of any country or territory.

Published in 1986 by the United Nations
Educational, Scientific and Cultural Organization,
7 Place de Fontenoy, 75700 Paris
Composed by Coupé, 44880 Sautron
Printed by Presses Universitaires de France, Vendôme

ISBN 92-3-102454-X

Preface

In the course of the last fifteen years there have been remarkable advances in two particular areas of science and technology: microelectronics, and molecular and cell biology. Discoveries in these fields have had repercussions on many other branches of knowledge and aspects of daily life, especially in the industrialized countries. The changes have been so radical and far-reaching that they have been characterized by the catch phrases 'information revolution' and 'biorevolution', terms which have become hackneyed, but which nevertheless reflect the perception that the world can never be the same again.

Thanks to microelectronics, computers have become more widely available and easier to use; they can accomplish increasingly sophisticated calculations and so permit processing and analysis of data on an unprecedented scale, thus facilitating research of all kinds. The applications of microelectronics seem potentially limitless, creating new products and methods, and altering the basis of many traditional activities. Virtually every aspect of political, economic and cultural life has been or could be affected. The impact of the biorevolution has been less conspicuous to the general public, and the applications are only beginning to be exploited on a large scale; the repercussions on human health, agriculture, food and energy production nevertheless promise to be as fundamental and pervasive as the information technologies.

These developments have taken place within the context of another major shift in the global situation, the sudden large increases in oil prices in the 1970s. Energy supplies became the focus of world attention, as net importers of energy sought for alternatives to oil and tried to cut back their demand. Whatever future developments in oil supply may be, energy will continue to be a major preoccupation of all countries for years to come.

The long-term effects of these changes cannot be foreseen, especially given that progress continues to be rapid in scientific research and the energy situation is far from static. It is clear, however, that no country can afford to stand back passively and watch either the information revolution or the biorevolution, since every one will ultimately be affected, directly or indirectly. The options need to be considered and a suitable response formulated.

It is the purpose of this book to try to clarify the issues involved, by describing the new technologies and some of their applications, and by discussing the possible benefits and challenges that they offer.

'Science and technology as factors of change: impact of recent and foreseeable scientific and technological progress on the evolution of societies, especially in the developing countries' is a topic of obvious general concern that has been considered within the framework of Unesco's Major Programme I: 'Reflection on world problems and future-oriented studies'. The present book evolved out of a colloquium, held at Unesco Headquarters in Paris, from 4-7 December 1984, attended by twenty-three participants from sixteen member states. Sixteen working papers were produced for the conference, covering both general trends (social change and development and adaptation of technologies) and specific applications (manufacturing, information technologies and telecommunications, food and agriculture, health, education and energy).

Each paper discussed the array of technologies involved and the related scientific research areas; the direct and indirect impacts on the evolution of societies, emphasizing the socio-cultural aspects; the forecasting methods and trends where appropriate; and sought to identify the issues and priorities raised by the technologies and their likely developments. The wide range of experience of the assembled experts ensured an extremely broad geographical perspective.

Following the conference, the contributed papers were drawn together in the light of the discussions and of the most recent scholarship. The coverage is, unavoidably, uneven. For some topics (such as information technologies and new manufacturing systems), there is as yet relatively sparse information about impacts on the developing world, but very abundant sources on the industrialized nations. On the other hand, the chapter on health concentrates almost exclusively on the developing world and deliberately ignores the remarkable progress in high-technology medicine, which is now available in most industrialized countries, but which is largely irrelevant to the needs of the rest of the world's population. Many of the themes overlap, and the division into separate chapters is to some extent arbitrary (information technologies are clearly fundamental to developments in education, for example). Each topic arguably merited discussion at far greater length, but at the risk of making the book unwieldy. For the same reason, it seemed more useful to provide national experiences drawn from across the globe in illustration of particular points, rather than to offer in-depth case studies of a few countries' experiences across the whole spectrum of new technologies. The references at the end of each chapter and the general bibliography indicate more detailed treatments in specialist publications.

The colloquium and the book would not have been possible without the invaluable assistance of all the participants whose names are given on pp. 11-12. Unesco is very grateful for the advice of Professors H. Brooks, A. Rahman and J.-J. Salomon, who took part in the preliminary working party to establish the framework for the colloquium and who have maintained a close interest throughout. The Organization owes special thanks to the authors who prepared the final drafts of the chapters, drawing upon the contributions of the colleagues listed in full in the Acknowledgements: Introduction (J.-J. Salomon), information technologies (H. Brooks, A. Lebeau, J.-J. Salomon), agriculture and food (M. Kenney, A. Sasson), health (G.S. Omenn), energy (U. Colombo), and education (R. Gwyn). Ann Johnston and Albert Sasson were responsible for the final editing of the manuscript.

This publication is the first number of a series entitled "Notebooks on World Problems". The aim of this collection is to circulate the most significant studies undertaken in the framework of Major Programme I "Reflection on world problems and future-oriented studies".

Future numbers in the series will focus on specific themes.

Contents

Acknowledgements

Unesco is deeply grateful to all those who contributed to this project, whether as authors of chapters and position papers or as participants in the colloquium. The following names need to be mentioned (the number preceding the name refers to the number of the chapter to which the contribution was directed).

2. Mrs P. Arriaga, Centro de estudios económicos y sociales del Tercer Mundo (CEESTEM), Magnolia 39, San Jeronimo Lidice, C.P. 10200 Mexico D.F. (Mexico).
2. Professor H. Brooks, Division of Applied Sciences, Aiken Computation Lab. 226, Harvard University, Cambridge, Massachusetts 01238 (USA).
5. Dr U. Colombo, Comitato Nazionale per la ricerca e per lo sviluppo dell'Energia Nucleare e delle Energie Alternative (ENEA), Viale Regina Margherita 125, 00198 Rome (Italy).
1. Dr J. Farkas, Institute of Sociology, Hungarian Academy of Sciences, 1014 Budapest 1, URI UTCA 49 (Hungary).

   Mr G. Ferné, Directorate for Science, Technology and Industry, Organisation for Economic Co-operation and Development (OECD), 2 rue André Pascal, 75775 Paris Cedex 16 (France).
6. Mr R. Gwyn, Didsbury School of Education, Manchester Polytechnic, 799 Wilmslow Road, Manchester M20 8RR (United Kingdom).
3. Professor M. Kenney, Department of Agricultural Economics and Rural Sociology, Ohio State University, 2120 Fyffe Road, Columbus, Ohio 43210-1099 (USA).
2. Mrs B. Krevitt Eres, Interdisciplinary Center for Technological Analysis and Forecasting (ICTAF), Tel-Aviv University, Ramat-Aviv, 69978 Tel-Aviv (Israel).

   Professor G. Kröber, Institut für Theorie, Geschichte und Organisation der Wissenschaft der AdW der DDR, Prenzlauer Promenade 149/152, 1100 Berlin (German Democratic Republic).
2. Professor A. Lebeau, Conservatoire National des Arts et Métiers (CNAM) (Techniques et programmes spatiaux), 292 rue St Martin, 75003 Paris (France).

   Dr A. Lemma, Technology Alert System (ATAS), United Nations Centre for

Science and Technology for Development, United Nations, New York, N.Y. 10017 (USA).

5. Dr J.D. de Melo Teles, SERPRO, Serviço Federal de Processamento de Dados, SGAN-QUADRA 601 - Modulo "V", CEP 70830, Brasilia, D.F. (Brazil).

4. Professor G.S. Omenn, School of Public Health and Community Medicine, University of Washington, Seattle, Washington 98195 (USA).

1. Professor K. Oshima, Technova Inc., Fukoku Seimei Bldg, 13F, 2-2-2 Uchisaiwai-cho, Chiyoda-ku, Tokyo 100 (Japan).

3. Dr B.N. Okigbo, International Institute of Tropical Agriculture (IITA), P.M.B. 5320, Oyo Road, Ibadan (Nigeria).

Mrs F.Z. Oufriha, Institut des Sciences économiques d'Alger, 2 rue Didouche Mourad, Algiers (Algeria).

1. Dr A. Rahman, Tower A, Flat 30, Zakir Bagh, Okhla Road, New Delhi 110025 (India).

4. Dr K. Raj, 84 rue Vergniaud, 75013 Paris (France).

1,2. Professor J.-J. Salomon, Centre Science, Technologie et Société, Conservatoire National des Arts et Métiers (CNAM), 292 rue St Martin, 75003 Paris (France).

4. Dr J. Serrano, Centro de estudios economicos y sociales del Tercer Mundo (CEESTEM), Magnolia 39, San Jeronimo Lidice, C.P. 10200, Mexico D.F. (Mexico).

2. Professor J. Sigurdson, Research Policy Institute, University of Lund, Magle Stora Kyrkogata 12B, Box 2017, S-220 02 Lund (Sweden).

2. Professor K. Valaskakis, Gamma Institute, 3764 Côte des Neiges, Montréal, Québec H3H 1V6 (Canada).

Mr V.N. Vasiliev, Department of Planning of the USSR Scientific and Technological Relations with Foreign Countries, 11 Gorky Street, Moscow (USSR).

5. Mr H.J. Wang, Standing Committee of the Techno-Economic Research Center, The State Council of China (TERC), Beijing (People's Republic of China).

Sir B. Williams, The Technical Change Center, 114 Cromwell Road, London SW7 4ES (United Kingdom).

1. Dr A.B. Zahlan, 74 Oakwood Court, Abbotsbury Road, London W14 8JF (United Kingdom).

1

Introduction: science, technology and society

'When I first formed the idea, about 1938, of writing a systematic, objective, and authoritative treatise on the history of science, scientific thought, and technology in the Chinese-culture area, I regarded the essential problem as that of why *modern* science (as we know it since the +17th century, the time of Galileo) had not developed in Chinese civilization (or in Indian) but only in Europe? As the years went by, and as I began to find out something at last about Chinese science and society, I came to realize that there is a second question at least equally important, namely why, between the −1st century and the +15th century, Chinese civilization was much more efficient than occidental in applying human natural knowledge to practical human needs?

The answer to such questions lies, I now believe, primarily in the social, intellectual and economic structures of the different civilizations' (Needham, 1964:127).

Joseph Needham's life's work shows us that certain societies, certain cultures, at various periods in history, reveal themselves as far more efficient than others in the mastery of scientific knowledge and the exploitation of technical progress. But it is not only the past that tells us this. At this very moment, even as there is talk of a new stage in the history of the Industrial Revolution with the advent of the 'new technologies' — information technology, telecommunications, biotechnologies, new materials (alloys, ceramics, composites), and a host of new products and processes which depend closely upon the most advanced work in both the laboratory and the factory and which at the same time changes the whole structure of the economic and social systems — it is clear to all that considerable disparities exist in the ability of different societies to take advantage of the possibilities opening up and, *a fortiori*, in their capacity to contribute to the conception, development and production of these new technologies.

In fact, matters are even more complex than Needham suggested in the article quoted above, which now dates back almost a quarter of a century. It goes without saying that, as Needham notes, one has to be 'deeply sceptical of the validity of any of those "physical-anthropological" or "racial-spiritual" factors that have satisfied a good many people'. It is more revealing, but also more difficult, to try to understand the *reciprocal* influence of science and technology on society: if the social and cultural factors — the attitudes and the beliefs attached to economic, political and social organization — influence the role that science and technology play in a given society, the spread of new knowledge, products and processes derived from scientific and technological progress in their turn transform social structures, modes of behaviour and attitudes of mind.

The role of technical change in the process of economic growth is recognized by all theories of development. But what precisely is that role? In particular, what part did science and technology play in the economic and social transformations that accompanied the Industrial Revolution from its beginnings? Answers to these questions can be neither easy nor, consequently, swift, requiring as they do a subtle analysis, a historical perspective, and reference to examples drawn from many different social sciences.

One thing is nevertheless clear. Today — even less than previously with coal and steel, then electricity supply, oil and the internal combustion engine — the ways in which technical change transforms attitudes, institutions and societies cannot be reduced to a simple linear relationship that is automatic, i.e. deterministic.

'Technology is one social process among others: it is not a question of technical development on the one hand and social development on the other, as if these were two entirely different worlds or processes. Society is shaped by technical change which, in turn, is shaped by society. Conceived by man... (technology) eludes his control only in so far as he wants it to. In this sense, society is defined no less by those technologies that it is capable of creating than by those it chooses to use and develop in preference to others' (Salomon, 1981).

At this very moment, therefore, and in a situation very different from the expansion of mechanization encouraged by the development of machine tools and the steam engine in the nineteenth century, the spread of the 'new technologies' creates far greater disparities than those that were possible between European countries at the beginning of the Industrial Revolution. Moreover, it involves much greater challenges than those tackled by nineteenth-century European societies (which were pre-industrial rather than purely agricultural), with ultimate success thanks to their long preparation in basing their interpretation of natural phenomena and their handling of techniques on, among other things, mathematics and experimentation — measurement, calculation and proof.

On the one hand, in fact, the geo-political situation in the world nowadays is more complex, with a flow of events and actors constantly in motion on a continental scale, further augmented by the explosion in the means of communication themselves. On the other hand, the very tools (both conceptual and practical) which allow us, at least partially, to understand the world in which we live and to manipulate it have continued — in large measure thanks to the spectacular progress of science and technology — to become ever more 'sophisticated' and therefore difficult to master without specialist skills and qualifications. Furthermore, one has to take into account the increasing unpredictability of economic factors such as commodity prices which may affect the long-term orientations of the national R&D effort. The 1973 increase in oil prices led to more R&D investment in energy alternatives; now that the cost of crude oil has dropped again, these research programmes have been cut back, if not abandoned.

It is against this background of increasing complexity of problems as much as of methods that the 'shock' of the new technologies has struck developing countries along with the industrialized countries. For these latter — after the economic difficulties of the early 1980s, the very moderate rates of growth and the persistence of high unemployment — the adjustment to the new technical system that is just beginning to spread poses problems that are not very different from those that gave rise to the various stages of mechanization of labour in the course of the nineteenth century. Whatever the social costs in terms of redundancies and job displacement, and however substantial the pockets of poverty that remain (and which sometimes even grow as a result of the crisis), we are nevertheless dealing with societies where basic needs are by and large satisfied, and furthermore the resources available to train and retrain the labour force are considerable. It is not for nothing that these have been called 'post-industrial' societies, characterized by the dominance of the service sector, the very rapid growth of information-related activities, and the large scale of investment in education and research.

By contrast, for most of the developing countries, the most basic needs for survival are far from having been met — food, shelter, health and education — so that the things that are perceived by the rich countries as essential can seem to the poorer countries like a display of luxury or a gimmick of the consumer society. In addition they face the double pressures of the population problem, which seems unlikely to see major improvement before the end of the century, and the debt problem, which has become so dramatic that some countries can barely cope with payment of the interest charges.

Against this background it is far from clear that some of the new technologies are what many developing countries want as a high priority in order to meet their real needs as speedily and as efficiently as possible. Yet at the same time — given both the growing inter-dependence of economies and the internationalization of trade on the one hand, and on the other the undeniable opportunities to modernize and 'catch up' that are offered by the new technologies — it is inconceivable that any country should choose to deprive itself of the products and the infrastructures that increasingly define the 'nervous system' of the contemporary world and determine its functioning.

The situation has been presented somewhat starkly, and clearly requires greater subtlety, or perhaps it would be better if all generalizations were avoided — for two good reasons. First, technology is never a single or isolated factor in the process of economic and social development, and the success of efforts at transplanting a technology depend to a great extent on the preparations made beforehand. Secondly, the similarity of the problems faced by developing countries should not obscure the great diversity of circumstances. This must be the point of departure for any understanding of the problems and the implications raised by the impact of the new technologies on the developing countries.

Technology in society

The rapid spread of a new technology does not of itself imply rapid social change. Other factors are involved, such as economic, social and educational policies, the negotiations and agreements between interest groups, the well-established customs of daily life and social institutions, the society's values and traditions. In the simplest terms, the societal impact of a technology depends on four groups of factors:

Scientific and technological. These factors are closely linked to developments in the laboratory, public or private, and are ultimately limited by the abilities of researchers, the quality of the equipment, the flexibility of organizational structures and the scale of financial investments. It is, however, essential to distinguish here between truly basic research, which aims to increase knowledge in general rather than to generate immediate results, as against all other forms of applied research and development that are intended to produce useful results in the short or medium term (even if the nature and the range of the final outcome are often very hard to foresee). Furthermore, a distinction should be made between R&D proper — which leads to discoveries and inventions — and technical innovation, which involves the launching of new ideas, products and processes onto the market. Scientific research often plays a lesser role in the success of an innovation than do ideas about management, design, production, marketing and advertising (Sappho, 1971).

Economic and industrial. Discovery and invention are the province of the researcher, innovation of the entrepreneur and the financier. Economic and industrial factors are thus the major element in a society's capacity for innovation. A bad or uncertain economic situation, a lack of knowledge or information, a lack of capital or the reluctance of investors to take risks, a lack of qualified manpower or the existence of stocks of existing equipment in firms and offices which cannot profitably be replaced — all these are factors that can prevent or delay the success of a new product or process. Clearly there are economic and fiscal policies that are more favourable than others to the encouragement of technical innovation, industrial structures more suitable than others to take advantage of technical changes in competition in national and international markets, social circumstances and attitudes more open than others to the notion of risk-taking.

Social. Even when an innovation seems to be economically advantageous for a firm or a country, its introduction may be delayed or ruled out as a result of the values and behaviour of potential users. The thinking of inventors and producers does not necessarily jibe with the ideas of consumers and users. New technologies need to be absorbed, and therefore require time and appropriate education if they are to spread through the society without too much resistance. The spread of micro-computers, for instance, is clearly linked to the mastery of the programs that come with them, and consequently to the policies of education and retraining at the level of both the firm and the nation.

Moreover, the adoption of a new technology may involve costs and risks that are often borne by only a fraction of the total population (working in mines, for example), while the benefits are widely distributed (growth of energy resources). This imbalance between the costs and risks can also work the other way, as for example when polluting industries have overall drawbacks that outweigh the local or individual advantages (to the industrialist, and also to the workforce concerned to maintain its employment). Who can have forgotten Seveso or Bhopal? In the name of the collective interest, the regulation of technical change clearly involves public scrutiny (as for new drugs, for instance) and the limitation of competition to prevent some innovations from showing greater drawbacks than advantages. It is all the more reason to increase scientific literacy as far as practicable in the population at large. Whether it be the recent European ban on steroid hormones in animal husbandry, the after-effects of the Chernobyl accident on different countries' energy programmes, or the pharmaceutical industry's tendency to test its products in regions of the world where they are less likely to encounter opposition, the influence of public (and thus political) opinion can be very important. In considering this factor, it must be recognized that a plea for greater public appreciation of scientific ideas is not a plea for greater public acquiescence; on the contrary, more widespread understanding may catalyse well-informed resistance to proposed new developments.

Institutional and cultural. Institutions, regulations and collective beliefs exist in order to ensure a certain equilibrium in the social distribution of benefits and risks. But innovation is by definition something that challenges habits, received ideas and traditional values. Some regulations may run counter to the spread of a new technical system (such as cable television networks). Other technologies may upset the beliefs and values of a group or a whole society (the contraceptive pill, for example). The degree of receptivity or resistance to the introduction of new technologies depends on the nature of those technologies and of the socio-cultural environment in which they seek to establish themselves, as well as the timing of their introduction — just as there are innovations that are not ripe for a particular community or society, so there are communities and societies that are not ready for a particular technical innovation, though the situation may change in the course of time.

Each of these four groups of factors should be studied carefully in any attempt to evaluate the societal impact of a new technology. But it is even more difficult to try to understand the way in which each interacts with the others in order to be able to analyse, let alone anticipate, the rate and extent of diffusion of an innovation, a task made even more complex in that it is a dynamic process and account should be taken of the socio-cultural circumstances in which the process takes place. In some industrialized countries, for instance, video systems were technically feasible but their launch failed because institutional factors came into play before the economic and social factors had been properly understood and adjusted for. Another example is the resistance of environmentalist groups in the United States of America or Austria to the construction of new nuclear facilities, even though in the 1970s the economic and industrial factors strongly favoured the growth of this form of energy.

All in all, there is no inevitability in technical change: neither its pace nor its direction is predetermined (even if it is unwise to underestimate the strength of certain industrial lobbies in imposing their factories or products), and the success of an innovation is never certain. Technology influences economics and history, but it is itself the product and the expression of culture — the same innovations can therefore produce very different results in different settings, or at different periods within the same society. Technical change and technology itself thus make up a social process in which individuals and groups always make the determining choices in the allocation of scarce resources, an allocation that inevitably reflects the prevailing value system (Rosenberg, 1979).

It is, therefore, easy to appreciate the limits and sometimes the failures of certain experiments in modernization conducted at headlong speed without regard for the economic, social or cultural realities of the societies into which they were being introduced; the utilization of science and technology cannot be reduced to the insertion of knowledge or know-how, techniques and methods into a social fabric that is unprepared. This fact underlies the equivocal nature (for some the illusion) of the notion of 'technology transfer', a transfer that involves much more than the movement of a physical object from one place to another. Transfers of technology require the preparation of education, management and production structures appropriate to the mastery of production of knowledge and know-how themselves (Salomon-Bayet, 1984).

Do such structures have to be identical to those that produced modern science in Western countries? Not necessarily, given the example of Japan, where the Meiji 'Restoration' led to the political decision to import the European scientific and technical model; the initiation into, and the rapid mastery of, Western scientific thinking came about not in terms of a rejection of a Japanese approach, but rather as its fulfilment. It is obvious that Japan never tried to follow the West blindly; instead, she tried to incorporate into her own system only those elements that would be of advantage in her task of modernization. This prudent and selective process of learning is often referred to as *wakon yosai*, meaning 'Japanese spirit and Western learning' (Hayashi, 1980). What distinguishes Japan from the European speculative heritage that dates back to Ancient Greece, therefore, is this attitude to science defined more in terms of its ability to produce practical applications rather than of its purely scientific creative power (Nagayama, 1983).

From this viewpoint, the 'universality' of science is illusory. The international network of scientists trained in the same institutions of higher learning and research, speaking the same language and publishing in the same journals, meeting one another periodically in the same places for colloquia and conferences, is indeed based upon the shared language, methods and results of a universal scientific community in the Western sense. For a researcher, the notion of belonging to the extended community of science is highly significant and supportive. But this international network of science, in much the same way as the airline routes, is not universal in the sense that absolutely everyone can join in: belonging to the network is not the same as sharing in the conceptual framework that gave rise to that network. An Indian researcher may share in the universality of science through having been trained in the methods of Western science and as a result may feel closer to his colleagues at the Sorbonne, Cambridge or the Massachusetts Institute of Technology than to his rural compatriots; but the same man may regularly consult astrologers, which his Western colleagues would consider to be a total denial of Western rationality.

The existence of the international scientific community means simply that if laboratories are extended like airline routes, scientific ideas can be moved about like aeroplanes. But planes do not land just anywhere – they need an infrastructure, which involves not only physical runways but also systems of maintenance and air-traffic control based on highly specialized qualifications. Moreover, planes are only built and flown when certain conditions have been satisfied and a price paid: as will be argued below (see Chapter 2), *mastery of production* and *mastery of use* of the new technologies do not necessarily go hand-in-hand. Finally, even if one international airport looks much like another because of a similarity of functions and architectural styles (to the point that their hotels give the impression of being simultaneously somewhere and nowhere), the traveller leaves the network as soon as he reaches the city proper, to be reabsorbed into the characteristic features of the individual national culture (Latour, 1982, 1983).

In other words, it is not enough to rely upon the universal methods of science and technology in order to reproduce a model of development based on a tradition, history and reality totally alien to that of most developing countries. What has been written about India in the aftermath of Independence is equally applicable to many other cases: 'Science has grown as an oasis in an environment which, if not antagonistic, is also not sympathetic to it, with the majority of people steeped in superstitions and traditionalism of which many of the leading scientists are also victims' (Rahman, 1977).

Hence the importance, of which the countries of the developing world have themselves become increasingly conscious in the course of the last decade, of development strategies that take care not to make science and technology a new source of disruption in the social fabric, with the scientists and engineers often becoming an elite whose interests are far removed from the economic and social preoccupations of their own countries. 'In a broad anthropological sense, technology is perhaps the most important element in any culture since, determining as it does the relationship between a community and its environment, it is thus the concrete expression of its values. Consequently one of the major aims of any development process in a poor country must be to put technology in its place as one of the crucial elements in its own cultural life' (Herrera, 1978).

The new techno-economic paradigm

A new technical system (Gille, 1978), a new techno-economic paradigm (Freeman and Perez, 1986), is currently evolving with various components that are both consistent with one another and interdependent. Up until the middle of the twentieth century, the major technical innovations of the Industrial Revolution arose out of knowledge that was generally available rather than drawn from the frontiers of science. Since the era of nuclear power and plastics, the new technical system is increasingly defined by innovations based to a great extent on very sophisticated science: space technologies, electronics, biotechnologies, new materials (ceramics, reinforced plastics, super-alloys, etc.). In comparison with all previous technical systems, the major characteristic features of the new one can be summarized as: a much greater complexity of conception, close linkages with scientific institutions, greater capital intensity, more diversified location of production, greater flexibility in application and uses, more rapid achievement of global development and world markets.

The situation has been clearly analysed by Freeman and Perez:

'The technological regime, which predominated in the post-war boom, was one based on low-cost oil and energy-intensive materials, (especially petrochemicals and synthetics), and was led by giant oil, chemical, automobile and other mass durable goods producers. Its "ideal" type of productive organization at the plant level was the continuous flow assembly line turning out massive quantities of identical units. The "ideal" type of firm was the "corporation" with a separate and complex hierarchical managerial and administrative structure, including in-house R&D and operating in oligopolistic markets in which advertising and marketing activities played a major role. It required large numbers of middle range skills in both the blue and white collar areas, leading to a characteristic pattern of occupations and income distribution. The massive expansion of the market for consumer durables was facilitated by this pattern, as well as by social changes and adaptation of the financial system, which permitted the growth of "hire-purchase" and other types of consumer credit. The paradigm required a vast infrastructural network of motorways, service stations, oil and petrol distribution systems, which was promoted by public investment on a large scale already in the 1930s, but more massively in the post-war period. Both civil and military expenditures of governments played a very important part in stimulating aggregate demand.

'The "ideal" information-intensive productive organisation now increasingly links design, management, production and marketing into one integrated system — a process which may be described as "systemation" and goes far beyond the earlier concepts of mechanisation and automation. Firms organised on this new basis... can produce a flexible and rapidly changing mix of products and services. Growth tends increasingly to be led by the electronics and information sectors, taking advantage of the growing externalities provided by an all-encompassing telecommunications infrastructure, which will ultimately bring down to extremely low levels the costs of transmitting very large quantities of information all over the world.

'The skill profile associated with the new techno-economic paradigm appears to change from the concentration on middle range craft and supervisory skills to increasingly high and low range qualifications and from narrow specialisation to broader multi-purpose basic skills for information handling. Diversity and flexibility at all levels substitute for homogeneity and dedicated systems' (Freeman and Perez, 1986).

The new innovations that evolve from this new technical system cause the expansion of new industries, based directly on 'intellectual capital' as much for the design of their products as for their production and distribution.

Competition between these industries takes place at the world level, and their commercial future is often determined in battles over the definitions and adoption of standards, i.e. the often intangible scientific systems or codes such as are used for colour television and radio reception or for micro-computers. The most obvious implication of these developments for public policy is that these industries cannot be 'nationalized': their physical plant can be seized, but not the intangibles that are the source of their technical and commercial success.

A new concept has had to be articulated to take account of these very sophisticated technological products: 'high tech'. The National Science Foundation puts an industry into this category if its expenditure on R&D represents 10% of the value added per product. High tech thus comprises pharmaceuticals, office equipment, electrical and electronic equipment, aerospace, missiles and satellites. In 1983 the expenditure of these industries in the United States of America amounted to 23% of their revenues (as against 7.5% for industry as a whole), and research staff (scientists, engineers and technicians) made up 16.5% of their personnel (as against 6.3% elsewhere).

This definition is not entirely satisfactory since it is based, in fact, on statistical categories established before the advent of the latest technologies (computers, for example, are included in the same category as typewriters); and the level of aggregation is too high to be able to pick out the high tech areas within traditional industries (the automotive industry, for instance, remains in the 'low technology' category even though increasingly vehicles are designed by computers and produced by robots). Furthermore, producers should be distinguished from consumers (this is essential for telecommunications and the media), and products from processes (the application of computers to management has entirely altered the basis of the industrial way of life, just as the refinement of programs so as to deal with the needs of industrial production has led to the growing use of micro-processors in traditional sectors).

Of all the new technologies, microelectronics is the most highly developed and the most widely adopted. Two strong trends have emerged: products are designed to satisfy user needs ever more closely as these become more complex; while the combination of electronics and telecommunications is transforming all systems of communication.

In fact the importance of the electronics industries cannot be measured solely in terms of that sector's statistics since the resulting technologies are both structural and strategic (Ramses, 1986). The full extent of their economic, social and cultural implications will be discussed below (see Chapter 2); for the moment what needs to be emphasized is their dynamic effect on economic growth, as they create new products and processes, new techniques of production and management that totally transform the structure of the economy. In order to remain competitive, traditional sectors (like textiles, steel, automotive, banking and insurance) must increasingly adopt electronics-based methods; the high tech industries disrupt the patterns of production and trade, leading to the development of more efficient and versatile robots, and bringing about considerable growth in productivity and a reorganization of activities in both factories and offices.

Furthermore these industries play a strategic role, in that the countries that have substantial electronics sectors and are leaders in the field enjoy an often critical advantage: this is especially true for 'custom chips' (made-to-measure circuits for information systems), without which a country's innovative capacity may be impaired and hence its competitivity. These technologies are all the more strategic in view of the role they play in the building of new weapons systems. The defence-related R&D effort is both a cause and an effect of the technological breakthroughs and innovations which ensure the competitive positions of the most industrialized countries in the field of new high tech. And it is only one step from restrictions imposed on access by foreign competitors to the most up-to-date products for reasons of national security to restrictions on narrowly protectionist grounds — a move that is often made not only against political opponents but also against allies.

The design and implementation of new technologies require enormous capital resources, a close link between science and technology and therefore between universities and industry, as well as a large pool of highly qualified scientists, engineers and technicians. The minimum entry requirements for industrial research are rising all the time. The 'smaller' a country in terms not of physical size but of financial and highly trained human resources, the more limited the range of scientific fields and branches of industry where it can hope to operate successfully. An 'offensive' research policy — in order to get far ahead of competitors — is increasingly costly, all the more so since the creation of a substantial part of the new technologies is linked to military research and hence public authorities alone can take on the risks and expenditures. Most countries, even if industrialized, are condemned to 'defensive' policies only.

Military R&D can certainly appear as having a limiting, if not actually detrimental, effect on the general well-being of the civil economy. Nevertheless, it does act as a considerable stimulant to the training of researchers and for R&D activities in general, some of which have medium- or long-term spinoffs that contribute to the creation of new technologies, if only by raising the level of qualifications of the scientific and technical labour force.

The large size of the capital outlays involved explains why the new technologies have led to a tremendous concentration of R&D efforts at all levels, and gives some idea of the enormous gap that separates the *most* industrialized countries from all the developing countries *taken together* (statistics taken from OECD Science and Technology Indicators, 1984):

National level. Without counting the Soviet Union, the majority of the R&D activities take place in a tiny handful of countries in the OECD area — five countries (United States, Japan, Federal Republic of Germany, France, United Kingdom) account for 85% of the total effort in the OECD area measured in terms of expenditures and numbers of researchers. Certain categories of R&D are so expensive (space research, high energy physics, development of 'fifth generation' computers) that they are beyond the reach of even some of the most highly industrialized countries, hence the increasing tendency in Europe to develop cooperative research programmes at the international level (like the European Economic Community's ESPRIT programme, or Eureka at a broader European level) or between firms from different countries.

Institutional level. The American Department of Defense alone controls 10% of all the expenditure on R&D in the OECD area, i.e. the equivalent of the *joint* expenditures of countries such as Italy, Canada, the Netherlands, Sweden, Switzerland, Australia and Belgium (or the equivalent of the research budget of the Federal Republic of Germany). Ten public and private institutions (government agencies and multinational companies) spend almost one third of the total R&D budget of the OECD area; some private institutions — like General Motors or IBM, or certain large North American universities — have research budgets equal to a third of the national research budgets of countries such as Austria or Norway. Eighty per cent of R&D activities are concentrated in companies with over 10,000 employees.

Technical area. In the United States of America and the United Kingdom, half of public expenditure goes on defence; in France and Sweden the proportions are almost as large. Among the specific goals of policies aimed at research and innovation in the OECD area, defence and space programmes account for 44%; telecommunications, transport, energy and urban renewal 21%; agriculture and manufacturing 15%; health and social services 11%; and basic research 10%. The United States' spending on basic research alone in 1983 ($4.7 billion) was equivalent to half of the *total* R&D budget of countries such as France and the United Kingdom. Private sector R&D efforts are concentrated in manufacturing, particularly engineering (space, electrical and electronic, computers and machinery, vehicles and marine) and chemicals (petroleum and coal byproducts, plastics and medicines).

The entire industrial system is shaped and manipulated by the 'big spenders' on R&D in their competition for innovation. This alters the basis of international trade as well as the margin of manoeuvre of the less big, let alone the really small, as was shown by an OECD seminar on 'Science and Technology Policy and its Relation to Small Industrialized Countries'. Although many of these are among the most technologically advanced and industrialized countries in world terms, there are nevertheless within the group relatively more and less advanced countries in terms of resources, *per capita* income and welfare. The poorer OECD countries have even less in the way of resources to spend on R&D and innovation than the richer ones, and are therefore faced with an even larger number of fields where minimum entry costs will be a barrier. For example, according to an EEC report (1985), Greece, Spain and Portugal have increased their R&D spending on biotechnology, new materials and information technology, but expenditure is still so low that it is relatively unproductive. The same report underlines that lack of resources in the less developed European countries may have been exacerbated by obsolete research structures and bureaucratic obstacles to entrepreneurship and innovation. 'The less industrialized countries are concerned that, having missed one industrial revolution, they are not left behind by the next' (Walsh, 1986).

All things are relative, of course: the 'less developed' countries of the OECD area do not fall into the category of 'developing countries' as defined by the United Nations. Nevertheless, faced with the exceptional concentration of R&D efforts, even some industrialized countries wonder whether they are threatened with exclusion from the war zone where the battles of international trade are to be waged in future. It is all the more understandable that the developing countries — defined in terms above all of their shortage of capital, abundance of labour and scarcity of food — worry about their margin of manoeuvre relative to the impact of the new technologies, in controlling the importation or trade, identifying those technologies that their economies really need and the sectors that could really make use of them without compromising their efforts to establish a more balanced programme of economic and social development, especially in coping with the chronic problem of underemployment.

To echo Marx's famous exclamation about the weakness of the Olympian gods compared with the power of the institutions that were the incarnations in his eyes of the vastness and dynamism of nineteenth-century industrial capitalism, what is the nation-state beside the multinational companies such as IBM or ITT whose R&D budgets are larger than the public expenditures of many industrialized countries, and hence of most developing countries? It is in relation to these disparities that the links between science, technology and society raise issues in terms of economic and political strategies.

The diversity of circumstances

Science and technology are not independent variables in the process of development: they are part of a human, economic, social and cultural setting shaped by history. It is this setting above all which determines the chances of applying scientific knowledge that meets the real needs of the country. It is not the case that there are two systems — science and technology on one side and society on the other — held together by some magic. Rather, science and technology exist in a given society as a system that is more or less capable of osmosis, assimilation and innovation — or rejection — according to realities that are simultaneously material, historical, cultural and political.

The scarcities that are characteristic of underdevelopment in what is called 'the Third World' are not of the same severity everywhere. There is not in fact *one* Third World, there are many, and the levels of underdevelopment are no less varied than the levels of development. From direst poverty to real possibilities of 'catching up', the diversity and the differences count for very much more than the resemblances and affinities.

The potential capacity of these countries to master their development depends on the size and rate of increase of their populations and — inseparable from that — the resources, both natural and human, that will enable a growth in their income. From this angle, the following groupings of countries may be identified, based particularly on their political organization and the strategies adopted towards industrial and or agricultural development:

— those where the capital resources, participation in world trade and degree of industrialization have permitted 'take-off'

— those where demographic pressures and the low level of mobilization of natural and human resources jeopardize the chances of catching up

— the oil-producing countries, lying between these extremes; one group is subject to strong demographic pressures, the others have a population growth that is not out of line with their oil reserves.

In the course of the last twenty years there has been a definite improvement in the pace of industrialization in developing countries. Their exports to Western countries have grown by 15% per annum in real terms, though they still account only for a minute part of the market in industrialized countries, and the share of developing countries in total industrial value added is growing extremely slowly. On closer inspection it appears that four countries are alone responsible for that share: Brazil, Mexico, Argentina and India. Only five countries export more than a billion dollars worth of manufactured goods per annum (excluding non-ferrous metals and petroleum products): Hong Kong, the Republic of Korea, Yugoslavia, Mexico and Brazil. Tiny Hong Kong alone accounts for one-fifth of the exports of manufactured goods by developing countries.

Moreover, these exports are limited to a very small range of products: clothing, metal and mechanical goods, textiles, wood products and furniture, food, shoes. As regards more sophisticated products (motors, electrical equipment, spare parts, microelectronics and components), the major part in these developments has been played by the multinational companies and the holding companies that have 16,000 subsidiaries and branches in the same countries: Brazil, Mexico, India, Pakistan, Philippines, etc. Recently, since 1982-3, a slight growth has been noticeable in exports of manufactured goods from developing countries to the United States of America and, to a lesser extent, Europe. A substantial part of this growth is due to multinational firms and to products related to electronics, though the competitive position of some countries would appear to derive from all sectors.

Nonetheless underdevelopment is not diminishing. On the one hand, the class of developing countries that exports industrial products remains very small. On the other, success in many cases has been achieved at the price of massive indebtedness and of internal distortions that cause a dangerous exacerbation of the tendency towards dualism in these societies — a tiny minority of the population enjoying the benefits of growth while the majority lag far behind, living in poverty and malnutrition, if not famine conditions.

None of the scenarios for the future is optimistic about the ability of the majority of developing countries to implement a rapid and balanced process of industrialization. If, arbitrarily, $2,500 average per capita annual income is reckoned to be the 'threshold' of development, it is estimated that between now and the end of the century the countries joining the group of those that will have 'taken off' will make up 12% of the world population (increasing from 470 million to 760 million). At the other extreme, if poor countries are defined as those with average per capita income of $300 or less, the percentage of the world population below that level will fall from 32% to 28% (1,650 million compared with 1,280 million at present).

Between the areas of desperate poverty (southern Asia and tropical Africa) and those countries or regions that have some hope of improving their position (Latin America, eastern Asia), there are many degrees of underdevelopment. According to the most moderate scenarios, the forecasts suggest a typology that should be seen not as a strict classification so much as an indication of the most characteristic conditions.

Industrializing countries with a diversified economy. These can be divided into two clear subgroups: (1) two medium-sized Asian countries, the Republic of Korea and Taiwan, plus the two city-states of Hong Kong and Singapore; (2) the biggest Latin American countries, Brazil, Mexico and, to a lesser extent, Argentina. Unlike the countries of the first subgroup, those of the second also have certain features of underdevelopment — such as a population that is mostly very poor, backward regions and periods of actual famine — alongside an advanced industrial, scientific and industrial infrastructure which produces certain goods that can compete in world markets.

Countries beginning to become industrialized, such as Algeria, Venezuela, Malaysia, Philippines, Pakistan, Nigeria, Kenya, Ivory Coast. This group is much more diverse, but again falls into two subgroups: at one extreme those that could, in very favourable circumstances, become industrialized nations; at the other, those whose possibilities of escaping from underdevelopment by industrializing are much less certain. Algeria is an example of the first subgroup, Pakistan of the second. In most of these countries agriculture will remain a crucial activity, often much more significant than industry.

Countries whose growth potential basically depends on natural resources, again divisible into several subgroups. This group includes countries that have an important market position in one or more ores or primary products (Saudi Arabia and the other OPEC countries for petroleum; Zaire, Zambia, Chile, Peru for copper; Thailand, Malaysia, Bolivia and Indonesia for tin; Jamaica and Guinea for aluminium; Malaysia for rubber; Ghana for cocoa; etc.); it also includes countries that have too small a share of world markets in several products to be able to undertake rapid industrialization (Ethiopia, Tanzania, Paraguay, etc.).

Very poor countries that lack natural resources and have little prospect of industrialization, so that improvements in agriculture are all the more pressing than in the other groups.

Country-continents — India and China — clearly stand apart. Their regional differences are blurred by their size, political unity and scientific heritage, yet they include within them features of all the earlier groups. India is simultaneously a developing country in its vast rural areas and an important industrial nation, especially in manufacture of capital goods. China, as a result of its efforts to control population growth and stimulate agricultural production, is basically rural but with increasing per capita income, and has pockets of industrialization and high technology that are spreading. The journal *Nature* caught the contradictions of India in a special issue entitled 'Excellence in the Midst of Poverty' (12 April 1984). In a recent special issue on China, the same journal emphasized the importance henceforth of science (it is one of the 'Four Modernizations', with agriculture, industry and defence), although also the long lags before the efforts to catch up in science and technology can be expected to pay off (21 November 1985).

This typology, obviously, is derived from scenarios related to major trends. If these trends are unlikely to change significantly before the end of the century, it is still possible that some countries within each group will be able to influence their futures by their development strategies — one way or the other. In any case, it is clear that for most developing countries, agriculture rather than industry will remain the dominant activity. Consequently any strategy that favours industry at the expense of agricultural production threatens in most cases to accentuate the internal distortions, and hence will compromise the chances of re-establishing a balance between the rate of population growth and the available natural resources.

Last but not least, it must be emphasized — whatever the scenario — that the obstacles to development do not simply involve the availability of natural resources vis-à-vis demographic pressures; they also involve social organization and political systems. Natural resources and demographic pressures alone dictate the threshold of utter poverty, but the political system and the nature of social organization define the limits of a country's ability to mobilize its human and financial resources. Social and cultural imbalances, political instability, inadequacy of economic choices can slow down or halt entirely the process of development.

Purely economic classifications of developing countries give a somewhat distorted picture; they do not take adequate account of differences in development paths that arise from, among other things, the nature of the political and economic régimes, the level and diffusion of education, the industrial and academic base, the class structure and the size of the professional class, the degree of 'duality' between a vast rural population and a highly educated scientific elite; or, indeed, the split within a single society between the silent minority that looks to the West for inspiration and the majority that attacks Western rationalism on the grounds of its religious convictions. The picture could be endlessly refined: the elements that reject scientific ideas based on the Western tradition are obviously not the same everywhere.

It is clear, however, that the transfer and assimilation of this scientific culture raises very difficult problems. Science and technology are not 'black boxes' with principles and effects that leave unchanged the social structures of the societies that adopt them. They cannot be shipped like commodities: the process is never neutral, straightforward or permanent; it demands levels of skill and often also perseverance without which it constitutes a tool without a handle or a box of tricks without a key.

It is from this angle — as much, if not more so, than from the angle of the availability of natural resources alone — that the links between science, technology and society in developing countries should be addressed. Beyond a certain threshold of resources, capital accumulation is never by itself a guarantee of growth. On the contrary, it is first and foremost the organization of society — which in turn determines the organization of production — that allows a country to create and exploit its scientific and technical resources. These factors define the extent to which science and technology can operate to initiate and stimulate the process of development, and not *vice versa*. If science and technology are not external to this process, it is because they cannot themselves be either developed or used other than in a given economic and social framework. Extreme underdevelopment is in this sense the stage of development that puts no pressure on the social structure to become involved in scientific and technical research. And, lacking a favourable economic and social structure, even countries above this level may find themselves unable to take advantage of science and technology.

The statement of an Indian scientist shows clearly that it is not sufficient simply to resolve the problem of financial resources in order to adapt *ipso facto* the research system of a developing country to the needs of advanced science:

'It has been pointed out again and again, by many authors, that it is immensely difficult to organize research institutes in underdeveloped countries and to keep abreast with latest developments. However, the point I want to make here is that, if we tacitly assume that these are the only obstacles in the way of starting a new scientific tradition, and that science would start blossoming in an underdeveloped country as soon as efficiently organized institutions are set up with good libraries and adequate research facilities, then we would be missing a most vital aspect of the problem. Science is one of the profoundest forms of creative expression of the human mind. Unless we have human minds properly conditioned to create science, it is absurd to expect science to stream out of buildings, libraries and laboratories, however well-equipped they may be' (Choudhuri, 1985).

The obstacles

For all these reasons it is essential to take into account the obstacles and the impediments that science and technology encounter when applied to the struggle against underdevelopment. The range and the forms of the influence that the new technologies can have in development strategies are clearly linked to the efforts made to overcome these obstacles and impediments.

Underdeveloped countries undoubtedly measure themselves rather more against the developed countries than against one another, but that does not mean that the problems that they confront can be reduced to a simple gap to be bridged by imitating the growth path of the industrialized countries since the nineteenth century; neither is it sufficient merely to graft on Western science and technology in order to generate growth. The one is linked to the other: any notion of development that is not conceived in terms of a global social phenomenon will inevitably come up against the obstacles and the inertia that hinder the application of science and technology in societies outside the Western cultural tradition (Salomon, 1985).

Inequality of resources. We have seen above how concentrated the efforts are in R&D in the most industrialized countries. The distribution of resources devoted to R&D at the international level is inevitably extremely uneven — between industrialized countries, and even more so between them and developing countries. The developed economies (including the centrally planned economies) account for 94% of R&D expenditures and 89.4% of the researchers (scientists, engineers and technicians). The developing countries are left with 6% of the expenditure and 10.6% of the researchers. The graphs that follow (Figures 1 and 2) show this enormous gap all too clearly. Of course, statistics in this area should be treated with caution, since there is no *direct* relation between the sum invested in R&D and the rate of economic growth. On the contrary, a country like the United Kingdom has had very moderate rates of growth despite high R&D spending, whereas Italy, which has invested little in R&D, has enjoyed a high rate of growth. Nevertheless, even if a direct connection between the level of R&D effort and economic growth cannot be established, there is evidence that the level of productivity growth, balance of payments, export performance and rate of innovation are indeed related to the level of R&D spending. The situation is thus extremely complex. The results of a national scientific and technological effort cannot be measured in the same way as GNP; R&D statistics need to be qualified with non-quantitative data.

Inequality of results. The scientific and technical activities of the industrialized countries generate methods, products and processes intended primarily to meet the requirements of their own economies — they are not, on the whole, designed for the needs of developing countries, and may indeed be completely useless and sometimes even harmful. Furthermore the dominance of the

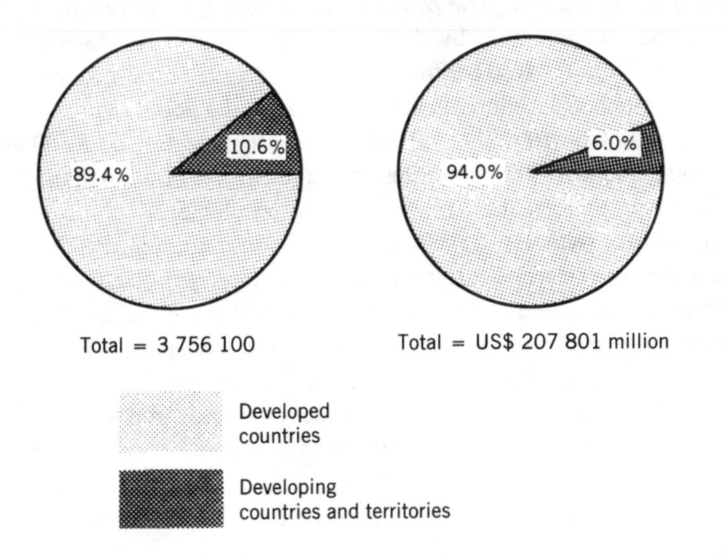

Figure 1. Distribution of scientists and engineers, and R&D expenditures (1980 estimates).

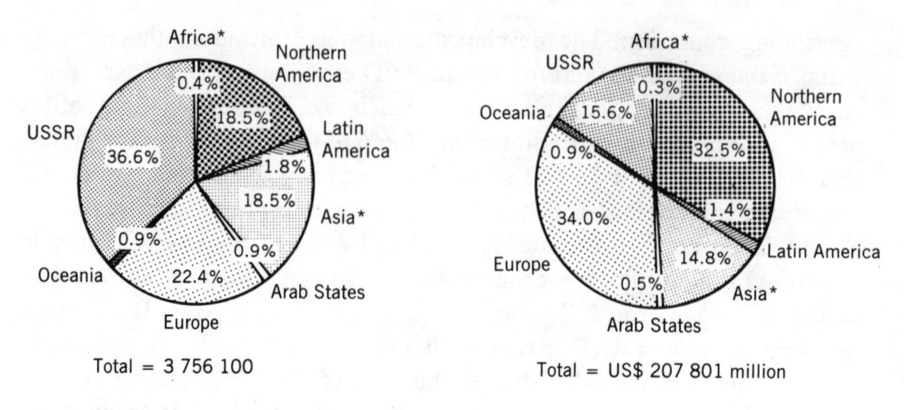

Figure 2. Distribution of scientists and engineers, and R&D expenditures by major regions (1980 percentage estimates).

military in the R&D spending of the most industrialized countries should be borne in mind. Internationally, 85.3% of patents are taken out by member countries of OECD, and of those six countries (United States of America, Japan, France, Canada, the Federal Republic of Germany, United Kingdom) account for 70.1%.

Limits of R&D research. The position of developing countries with regard to science and technology is by definition precarious in comparison with the highly concentrated R&D efforts of a few industrialized economies. India spends as much as Belgium on R&D, although the population is 600 million as against less than 10 million. Besides, the official figures on R&D efforts in developing countries are often exaggerated (estimates of 0.1% to 0.5%) for the simple reason that the actual figures are not available.

Above all, statistics do not take into consideration the fact that the organization of research is utterly unlike that in industrialized countries, even if certain countries and certain research institutions are the equals of the best Western universities. Research efforts are often dispersed among a large number of small centres or laboratories, the underfinancing and underequipment of which merely add to the weaknesses of the research work undertaken, whether in universities or government institutions — not to mention the administrative procedures and the bureaucratic controls peculiar to economic and political systems that cannot appreciate the real requirements of research work.

Relative costs of research. The paradox is that the countries in which the

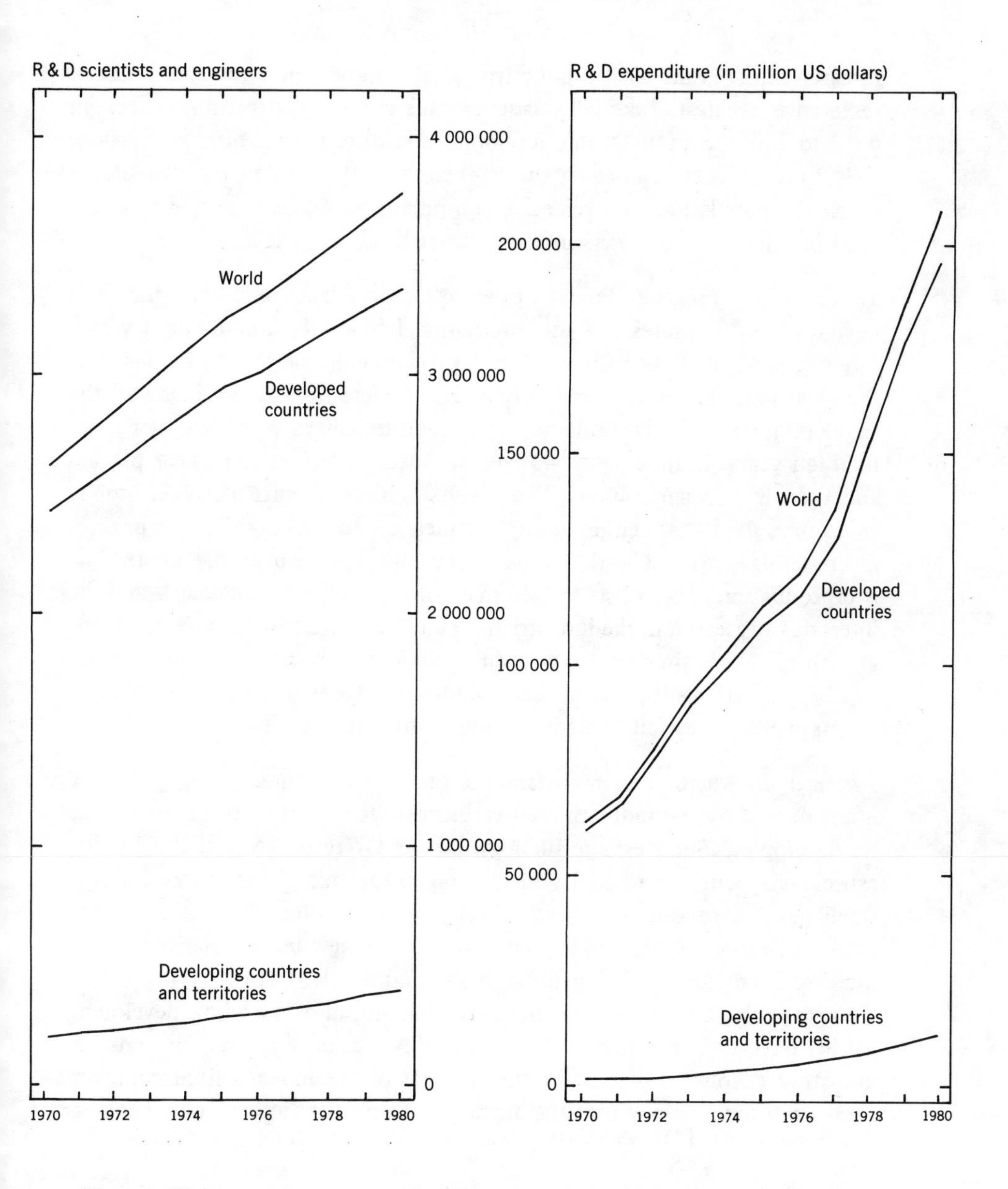

Figure 3. World trends in numbers of research scientists and engineers, and R&D expenditures (1970-80 estimates).

research effort is the least productive are also those where the social cost of research is greatest. Like all paradoxes, this is hardly surprising in fact: in order to limit the 'brain drain', developing countries have to offer proportionately much higher salaries to their research staffs than to the rest of the employed population, and scientific equipment has to be imported at prices fixed in relation to the costs in the industrialized countries.

The distortions of science. Research programmes in science tend to be drawn up in line with the interests of the international 'scientific community', which rarely coincide with the realities of the local economic and social situation. For the most part, biomedical and agricultural research on the problems of the developing countries is conducted by the industrialized countries. For more than ten years, India devoted 40% of its research budget to nuclear physics and only 8% to agriculture. The results achieved, particularly in atomic weaponry, show that neither ability nor means were lacking. But the priority given to this sector has by definition been at the expense of all the others. The overproduction of scientists trained in accordance with the standards and the interests of research in the industrialized countries contributes directly to the growth of the 'brain drain', while an inadequate number of institutions are capable of training technicians and middle management — poorly suited to their purpose, they fail to attract the best students.

Technical distortions. The importation of production techniques designed and implemented by the industrialized countries often involves three drawbacks for developing countries: a technology ill-suited to resources available locally, especially labour; overcapacity in relation to the size of the market; a very small gain in productivity across industry as a whole. For example, the industrialization of developing countries has generated a relatively small increase in employment. Argentina, Brazil and Mexico (which between them are responsible for 42% of the manufacturing output of the whole developing world) have certainly matched and sometimes even surpassed the ratio of industrial output to gross domestic product of the industrialized countries (37%), but their ratio of manufacturing employment to total employment has remained well below (22 to 29% as against 39%) (World Bank, 1980).

Limits of technology transfer. The dominance of the part played by the multinational companies in the process of technology transfer has been remarked so often that one hesitates to add to the chorus. Nevertheless it must be pointed out here that the bulk of technology transfer happens not through direct investment, but by flows within international companies (internal transfer). It is easy to understand why this type of transfer has caused so much discussion about problems of dependence and their implications for the economic and social fabric of developing countries. For one thing, the exported branches of research are mostly in medium or low technology sectors (i.e. most multinationals do not develop research in the latest and most sought-after technologies in their subsidiaries). For another, the weight of these firms is such that the introduction and implementation of new technologies results from decisions taken without regard to the political and economic context of developing countries. From this viewpoint — hence the ambivalence of recent debates about the role of such firms — the multinationals simultaneously contribute to the acceleration of industrial transformation and threaten to break up the existing balance between manufacturing sectors, if not the entire economic structure, of the receiving country (Madeuf, 1981).

Lags in the pace of innovation. The mastery or control of the pace of technology transfer is made all the more difficult for developing countries by the fact that, on the one hand, the pace of technical change continuously increases in some sectors, while on the other, such countries are often least well equipped to evaluate, choose, assimilate and adapt foreign technologies for themselves. The example of microelectronics — where the applications are spreading from the service sector into manufacturing methods — shows how vulnerable the industrial development strategies of developing countries are to the spurts of technical change in the industrialized countries. The rapid development of automated production by industrial robots is leading simultaneously to a halt in the policy of delocalization by firms in industrialized countries — which some developing countries were counting on to create employment — and to a growth in the economic and social difficulties of developing countries as the increases in productivity and profitability add to their problems of unemployment (see Chapter 2).

We shall see in Chapter 3 how numerous and diverse are (and increasingly will be) the opportunities opened up by the progress of molecular biology. Far-reaching changes are likely to stem from current developments in this field which promises great benefits for the less developed countries. Yet one has to be cautious as to the possible negative effect of the widespread application of biotechnologies. The hopes placed on the expansion of these new technologies carry the same threats of distortion as microelectronics which the developing countries will be the first to suffer. By reviving the pace of growth of the chemical and agro-food industries, the expansion of the biotechnologies may lead to a replacement of the products of poor countries by a new range of products able to sweep the international market. To add to these external 'perverse' effects, the repercussions of scientific progress in the industrialized countries, there may be effects brought about by the adoption of the most advanced biotechnologies by the developing countries themselves. If genetic engineering acts as the substrate of these new capital-intensive industries, there is a risk that these industries once again will run counter to the real needs of developing countries, which need to save not labour, but capital.

The question of the control or mastery of technology, which is also raised in the industrialized economies, is certainly much harder to resolve in the developing countries. Ultimately, the question goes back to the same problem of the links between technical change and social organization. It is all the more difficult for the developing countries to resolve, in that their situation of technical dependence is, in most cases, accompanied by economic and social structures that hardly lend themselves to internal use of scientific and technical resources — when such exist in sufficient quantities for research programmes to be directed towards local needs.

The need for a range of strategies

All the thinking about development since the Second World War has been influenced by the theory of 'take-off', even when it has involved criticizing the theory's weaknesses. Obviously, first for the economists, then for the historians, the idea of take-off has helped in understanding some of the elements in the Industrial Revolution, by clarifying the mechanisms that at a given moment altered the old rules of the economic game, leading to a very brief start-up period followed by accelerating growth.

Quite apart from the limits of the theory in explaining the causes of growth in different countries, it is obvious that the conceptual tool has been mishandled in applying it to contemporary developing countries, as if this model — possibly valid for the pre-industrial economies of Europe in the seventeenth and eighteenth centuries — had a universal truth. To do so is to overlook three factors that the theory of take-off has always played down, if not ignored completely:

– First, societies that are developing today, that have for the most part been subject to domination by the first nations to industrialize, do not have the same economic, social and cultural situation as the pre-industrial European countries. For instance, the sprawling capital cities of some developing countries, which double their shanty-town suburbs in under a decade, are growing at a pace and on a scale utterly unlike that of the European capitals in the course of their industrialization.

– Then, just as the theory omits to mention the relative strengths of resistance and progress in the evolution of Western societies, so too does it overlook the imbalances and distortions, and the internal crises that growth causes in developing countries. By giving pride of place to the 'positive' aspects of progress at the expense of traditional values, history has been made into a linear progression in which the process of modernization is necessarily a change for the 'better' (for example, Rostow's 'mature' stage). It is, however, this change composed of resistance, reversals and suffering that provides the context for history as people live and make it, not as the economists imagine it.

– Finally, in this definition, technical change has to be a given imposed from outside — no matter what the economic, social and cultural setting — and not a variable of which the influence, the ability to spread and be assimilated depend upon these very factors. This assumption is even less likely when, as was the case at the beginning and above all in the second half of the twentieth century, technical change is brought about by a close alliance of science, technology and industry — not as a technique developed independently from science (as in the days of the steam engine), but as a social process which requires the conjunction of highly abstract and sophisticated knowledge with a science-based technology and production methods that are ever more closely linked to 'intellectual capital'. In modern terminology this is precisely what 'technology' means: the conjunction of the laboratory, the university and the factory.

The theory of take-off with its conceptual approach would not have been important had it remained merely a subject of academic debate. But it must be acknowledged that these ideas have inspired, more or less directly, notions and policies of development on the part of governments as well as international organizations, to the point of forgetting the essential element: the *social* dimension, with its checks, resistances and blockages has been underrated in the interaction between science, technology and society. So, too, has the *time* factor in the transition between the pre-industrial stage and 'pseudo-maturity', as well as the capacity to take advantage of the knowledge and techniques available in the same conditions as those that the industrialized countries experienced.

One needs to abandon a Western ethnocentric viewpoint in order to begin to understand that the developing countries are not exact replicas of the pre-industrial European societies. And that, arising out of different circumstances, they *create* different circumstances. It is not sufficient simply to rely on science and technology to reproduce time and again the same model of development when the bases are so fundamentally alien to the realities of most developing countries.

Great stress has been put here on the diversity of circumstances in the developing world. Clearly, the contributions of science and technology to the development of the Third World have been already important and in some sectors decisive (health, agriculture, food), with the spread of the new technologies, more and better suited opportunities for such contributions will be available to increase and accelerate the process of development. The first conclusion to draw from the past experience is that the diversity of experiences should be matched by an equivalent range of different general strategies of development, especially in regard to the application of the resources of science and technology. Just as there is no single model of development that can be exported from the 'advanced' countries, so there is no single prescription applicable to all the non-industrialized countries.

In fact the debate is more vigorous than ever between those societies that are pinning their hopes on the transfer of the most advanced technologies in order to move out of underdevelopment and those that are relying on fairly 'traditional' technologies — more or less 'appropriate', more or less likely to create employment rather than capital. Should economic choices disregard social organization? Some say that technology cannot be made to suit developing countries; what adds to their social well-being and independence is the volume of goods produced, not the number of jobs created in order to produce that output. Technology that is not very advanced does not provide outlets for technicians; now, since multinational companies encourage, through their presence and their recruitment, the training of a specialized labour force, the shortcuts for the developing countries are mediated through these firms rather than through the modification of social structures (Emmanuel, 1980). Others doubt the beneficent role of the multinationals in helping the developing countries to catch up technologically, and they would attribute the benefits to the pressures and directions applied by the governments of the countries concerned (this is the view of Furtado and Elsenhans; in Emmanuel, 1980).

In response to the diversity of the developing world at least two kinds of strategy may be identified, though they could easily be complementary and adopted jointly in certain instances, depending on the particular circumstances; in others, however, they are clearly mutually exclusive.

At one extreme there is the classic approach that aims at growth in national income through increasing investment, industrialization and participation in world markets. For those countries that have achieved a certain level of development, the process of industrialization is thought to rest upon a Western-style university system, the training and employment of a highly qualified and specialized work force, and a complex organization of scientific research. A policy for science and technology that is basically inspired by those of industrialized countries is thought to help to back up the spread of knowledge and the application of techniques, as well as ensuring that the human infrastructure is there to deal with the transfer of the most advanced technologies which could not otherwise be assimilated under local conditions.

At the other extreme, development is seen in terms of covering basic needs rather than in terms of 'take-off'. For such countries it is obvious that not only is the classic approach inadequate, care should be taken to avoid even being influenced by it. Where agriculture is, and promises to remain, the dominant activity, it is not possible to talk about a policy for science and technology as if the same mechanisms, institutions, arrangements for training and dissemination of knowledge were involved as in a process of industrialization.

Between these two extremes it is possible to imagine mixed strategies that aimed simultaneously to strengthen and develop the purely scientific research structures and to support the expansion of primary activities (agriculture and extractive) through professional training based on local needs. The idea would be to increase the productivity of traditional technologies and also modify some of the advanced technologies to the most pressing general needs (Lucas, 1983; Weizäcker, 1983). Just as there are intermediate technologies, so there are intermediate levels of knowledge between the traditional and the scientific.

A third way could be that of technological blending, defined as the integration of newly emerging technologies with traditional modes of production and activities to ensure higher productivity while retaining some of the traditional characteristics of conventional techniques. Such an approach can also cover cases under which specific new technologies are integrated with traditional ones, e.g. microelectronics control for biogas production, and use of numerically controlled devices for lathes, etc. In cases such as the latter, the concept may involve unpacking different elements of a conventional technology and replacing some functions by microchips and numerically controlled units to ensure better control of product quality and supply of raw material inputs. This technique may be considered as an example of blending in the modern industrial sector (Bhalla, 1986).

The experience of failure of strategies based on the emulation of Western methods shows that it is not enough to possess a scientific elite in order to cope with the enormous challenges of underdevelopment. Has the mirage of success held up by the Western model made us forget the conditions that had to be fulfilled in Europe, the price that had to be paid, the elapse of time before reaching 'maturity'? People are dazzled by the entrepreneurs, the captains of industry, the big bankers and capitalists, the inventors and engineers who were responsible for the growth of industrialization, but they forget the social costs, the traumas, the distortions and crises that accompanied the whole period of emergence from the pre-industrial era. Furthermore, little attention is paid to a crucial factor in the expansion of industrial societies: the widespread possession of basic technical skills associated with the spread of primary education. All in all this is to forget that the first Industrial Revolution was less the product of science and scholars than of methods and foremen-craftsmen with a minimum of mathematics.

Similarly today, there is no alternative to a policy of technical education and research that tries to be as far as possible endogenous and suited to the economic and social conditions of each country.

'It is not sufficient to recognize the necessity of rational scientific research that on its own will allow a developing country to understand and therefore exploit the scientific result produced by the international community. This research must in addition be conducted in the country, by its nationals... Politically, having recourse to a scientific "Foreign Legion" tends to dissuade governments from making the effort necessary to provide themselves with their own research system. Scientific assistance exudes perverse effects. Given the role of science in development, it is conceivable to put it into the hands of scientific "mercenaries"... Finally, in economic terms, research elicits the appearance of nodes of creativity and imagination in the machinery of the State and in firms. Such a role of questioning, of positive challenging, is as useful as the innovative function (Pisani, 1984).

The price of adjusting to the impact of new technologies is thus an education policy directed at technical skills at *all* levels of the educational system. In the interaction between science, technology and society it is not science *per se* that plays a decisive role in the evolution of most developing countries. Well-trained technicians and middle managers can be more valuable than scientists with doctorates recognized by the international scientific community. However, this is not a reason to neglect the significance of the social base which influences the creation of conditions favourable to the progress and spread of technical expertise in agriculture, agro-industry, small business: the spread of scientific practices and methods starts first with the spread of a technical culture that takes account of the realities of that social base. And, from this angle, there can be no shortening of the time needed to disseminate widely a technical culture capable of increasing the productivity of even traditional technologies.

In any case, the contributions of scientific and technical research to the development of the Third World will depend, for a long while yet, upon cooperative research activities undertaken with the help of the industrialized countries and multinational organizations, such as the United Nations family. However, the cooperative research undertaken between the developing countries themselves — either on a regional basis or, whatever their geographical separation, as a function of common structures, levels of development and interests — will be increasingly important in the future.

References

BHALLA, A.; JAMES, D.; STEVENS, Y. 1986. *Blending of New and Traditional Technologies.* International Labour Organization and Dublin, Tycooly.

CHOUDHURI, A.R. 1985. Practising Western Science outside the West: Personal Observations on the Indian Scene. *Social Studies of Science* (London), Vol. 15.

EEC. 1985. *Report on the Differences in Technological Development between the Member States of the European Community*, European Parliament, Brussels, Working Document A2-106.

EMMANUEL, A. 1980. *Technologie appropriée ou technologie sous-développée?* Paris, Presses Universitaires de France.

FREEMAN, C.; PEREZ, C. 1986. The Diffusion of Technical Innovations and Changes of Techno-Economic Paradigm. Paper given at Conference on Innovation Diffusion. Cadolfin, Venice, 17-22 March 1986.

GILLE, B. 1978. *Histoire des techniques*. Paris, Gallimard, Pléiade.

HAYASHI, T. 1980. *Historical Background of Technology Transfer: Transformation and Development of Japan*. Tokyo, United Nations University.

HERRERA, A. 1978. Tecnologias cientificas y tradicionales en los paises en desarrollo, *Comercio Exterior* (Mexico), Vol. 28, No. 12.

LATOUR, B. 1982. Le centre et la périphérie : à propos du transfert de technologies, *Prospective et Santé* (Paris), No. 24 (Winter)

– 1983. Comment redistribuer le grand partage?, *Revue de Synthèse* (Paris), No. 110 (April-June).

LUCAS, B.; FRIEDMAN, S. (eds.). 1983. *Technological Choice and Change in Developing Countries: Internal and External Constraints*. Dublin, Tycooly.

MADEUF, B. 1981. *L'ordre technologique international*. Paris, La Documentation Française, Nos. 4641-2.

NAGAYAMA, S. 1983. The Transplantation of Modern Science to Japan. Kingston Conference on Science and Society, Unesco.

NEEDHAM, J. 1964. Science and Society in East and West. In: *The Science of Science (Tribute to J.D. Bernal)*, ed. M. Goldsmith and A. Mackay. London, Souvenir Press.

OECD. 1984. *Science and Technology Indicators — Resources devoted to R&D*. Paris.

PISANI, E. 1984. *La main à l'outil : le développement du Tiers Monde et l'Europe*. Paris, Laffont.

RAHMAN, A. 1977. *Triveni — Science, Democracy and Socialism*. Simla, Indian Institute of Advanced Study.

RAMSES. 1986. *Rapport annuel mondial sur le système économique et les stratégies*. Paris, Atlas-Economica.

ROSENBERG, N. 1979. Technology, Economy and Values. In: *The History and Philosophy of Technology*, ed. G. Buliarello and D.B. Doner. Chicago, University of Illinois Press.

SALOMON, J.-J. 1981. *Prométhée empêtré — La résistance au changement technique*. Paris, Pergamon (reprinted Anthropos, Paris 1983).

– 1985. La science ne garantit pas le développement, *Futuribles* (Paris) June.

SALOMON-BAYET, C. 1984. Modern Science and the Coexistence of Rationalities, *Diogenes*. (Paris), No. 126, p. 198.

SAPPHO PROJECT, 1971. *A Study of Success and Failure in Innovation*. Science Policy Research Unit, University of Sussex.

VON WEIZACKER, E.V.; SWAMINATHAN, N.S.; LEMMA, A. (eds.). 1983. *Integration of Emerging and Traditional Technologies*. Dublin, Tycooly.

WALSH, V. 1986. *Technology, Competitiveness and the Special Problems of Small Countries*. Background Paper submitted to the OECD-Finnish Government Seminar on Science and Technology Policy and its Relation to the Economic Growth of Small Industrialized Member Countries, Helsinki, 29-30 January 1986.

2

Information technologies and telecommunications

The information revolution

'Without materials nothing exists, without energy nothing happens, without information nothing makes sense' (Oettinger, 1984)

The most remarkable aspect of change in the industrialized countries in recent years has been the growth of the so-called 'information sector', i.e. that based on computers and the multiplicity of machines associated with them, from programmable robots to wordprocessors, plus the symbiotically related telecommunications technologies, from televisions to telephones to space satellites. Many commentators have argued that this marks a major and irreversible move towards the post-industrial, 'information society', first identified in the United States of America but now characteristic of the whole of the developed world (Bell, 1973; Porat, 1977; OECD, 1980). This information revolution has had a pervasive impact throughout the economies of the developed countries, affecting agriculture, mining, manufacturing and especially services. It has created new techniques, new products, new skills, new modes of work, education and leisure, and in the process has displaced many traditional products and activities.

Like earlier technical changes, the new information technologies have inspired both hopes and fears. The optimists emphasize the potential for growth and autonomy, the enhancement of skills and culture, the elimination of all unpleasant and boring tasks, whereas the pessimists fear dehumanized work, growing unemployment and bureaucracy, with unavoidable threats to privacy. Either way, there are likely to be profound social and intellectual upheavals as a result of the shift away from traditional forms of work and thought. No technical system has a predetermined result, however. Its effects are strongly influenced by the evolution of societies themselves; the final outcome is as much the result of the way that new technologies are adopted and implemented, in the local or global context, as it is of the strictly technical possibilities involved. There is all the more reason, therefore, to insist that the possible consequences of the new technologies for the developing countries are not necessarily going to be experienced in the same way as in the industrialized nations.

Evolution of the information technologies

In contrast to earlier periods of technical change which were linked fundamentally to matter and energy, contemporary change and growth are characterized by the predominant role of a third physical entity, information (Lebeau, 1972; Oettinger, 1984). This is not to say that matter and energy are no longer important or have ceased to grow, but their growth is more in terms of complexity than volume, a complexity that is closely bound up with the influence of information technologies. The increase in information transactions is the most characteristic phenomenon of contemporary civilization, and their level is closely correlated with the overall degree of development, adding an intangible element to the disparities reflected by the more traditional indicators, mastery of matter and energy.

Like matter and energy, information is measurable in terms of physical units (thanks to the work of Shannon and Weaver, 1949; Brillouin, 1956; and others); this lies at the heart of the theories and practice of information technologies. Information can be transferred, processed, stored and used in much the same way as are matter and energy. But in other ways the concept of information is quite unlike matter and energy. There are no physical barriers to its continued growth, in that it does not involve the exhaustion of non-renewable resources or the disturbance of the ecosphere; it is true that it has to operate within the limits of certain natural resources such as the electromagnetic spectrum and the geostationary orbit, but it makes use of them without using them up. Moreover, information has social dimensions in a way that matter and energy do not: in addition to meeting the direct needs of individuals, information transactions mostly arise from the individual's membership of a group (access to the knowledge that is required for group membership; information related to administration, provision of services, etc. for the group); as we shall see, information thus has great importance for the maintenance of a collective cultural, political and economic identity.

Are there limits to the astonishing growth of the information sector? The lack of physical constraints has already been mentioned. So far there do not appear to be technical or intellectual barriers to continued growth either, since the generic technologies involved are still far from reaching their theoretical potential; ultimately, the capacity of the human mind might appear to be the limiting factor, but this eventuality is still a long way off, and, in addition, fully automatic systems generate a volume of information transactions that bypass any limitation imposed by the human brain. Furthermore, supply is continuously stimulated by demand since it is associated with the increasing complexity of economies and societies, rather than merely with population size and levels of per capita consumption. The innate tendency seems to be towards ever increasing size and complexity, even if in theory there are alternative models of development.

CATEGORIES OF INFORMATION TRANSACTION

Five technical operations can be distinguished in all informations transactions, regardless of their content or purpose:

- — processing
- — transfer
- — storage
- — capture
- — retrieval

The first three take place *within* the technical system, while the other two involve movements into or out of the system to human or other destinations — in a form adapted to man if via a human operator (e.g. a micro-computer with its keyboard, screen, printer, 'mouse'; a television network with its microphones, cameras and the screens of the receiving sets), otherwise in a mechanical form made up of sensors that measure a physical parameter and actuators that act on the basis of the information received (e.g. a robot on an assembly line, or the automatic pilot of an aircraft), or some mixture of the two (e.g. the computer-assisted guidance system in a space shuttle, or the control system of an oil refinery or a nuclear power station that operates automatically but at the same time keeps human operators informed and therefore able to intervene if necessary).

Usually a technical system can be classified according to which of the first three operations is paramount: for instance, a computer is involved in data processing, a telecommunications system in information transfer, a data-bank or a library in storage. (The other operations may be essential but subordinate, as in the case of the computer in which internal storage and transfer facilities are required in order for it to function, and even determine the processing capacity of the machine.) The techniques required for capture and retrieval are ancillary to all three, thus the same screen VDU can be used to retrieve information from a computer, a television network or a telephone line.

By and large, technical systems do not themselves generate information and their limitations are well summed up by the old saying 'garbage in, garbage out'. Nonetheless progress is being made in the field of 'artificial intelligence', which attempts to simulate certain aspects of human intelligence. It is impossible to foresee the ultimate potential of machines capable of modifying their behaviour on the basis of past experience; at present, the practical consequences are still marginal, though this may prove to be a very transient situation.

'Generic technology' is the term used to describe the knowledge that underpins a particular technical field. It is at this level that the relationship is determined between research and development — between scientific and technical knowledge — while mastery of the generic technologies is the point of departure for further stages of technical progress. An appreciation of generic technologies provides a useful perspective on the development process in the industrialized countries and also helps in identifying some of the obstacles faced by the developing countries in their attempts to catch up.

EARLIER TECHNICAL DEVELOPMENTS

Technological progress in the field of information is not merely a recent phenomenon. The invention of movable type, making possible a massive dissemination of written information, signalled the start of modern times, though it was not until the nineteenth century that the two fundamental discoveries took place that mark the beginning of our present era: the transfer of information first by electric current with the invention of the electric telegraph by Samuel Morse in 1832, then by radio wave with Marconi's first transmissions in 1889. These two developments have been fundamental to subsequent technical progress that has increasingly relied upon the use of an intangible medium: electric current or electromagnetic waves. This process has resulted in a gradual 'dematerialization' of information that affects all sorts of data and every variety of transaction, as these methods steadily replace the alternative approach which involves the registering of information on some kind of substance (for example, in the substitution of telecommunications for the postal service). For long-term storage of information a tangible medium remains essential; this can be achieved without altering the form or chemical composition of the medium, but merely by making reversible changes in its physical state — by changing the magnetization of magnetic memories, for example. The general advantages of this trend are obviously the greater speed and the reduction in energy required for transactions.

As a consequence of these developments, the generic technologies relating to electron flows and electromagnetic fields have assumed special importance, above all electronics (strictly speaking, the manipulation of electron flows, though the term is often used more loosely to cover the range of technologies dealing with photon flows: lasers, fibre optics, etc.).

The connection between the information technologies and the generic technologies is extremely complex; the Japanese, who have been particularly successful in this area, like to conceive the linkage in terms of a bonzai tree, of which the roots are the generic technologies, the trunk is where these technologies come together, and the branches represent the different applications of these combined technologies.

In analysing the trends in technical progress it is in general useful to treat information transfer separately from processing and storage on the one hand, and capture and retrieval on the other. The trend towards miniaturization of electronic components, however, has affected all these aspects. The development of microelectronics dates back to the invention in 1947 of the transistor, which gradually replaced the thermoionic valve. The early 1960s mark the start of microelectronics proper, as several transistors began to be grouped together on the same semi-conductor chip (usually of silicon) with other parts of electric circuitry (capacitors, resistors) in an 'integrated circuit'; the 'microprocessor' appeared a decade later, with the gathering together of all the functions of an electronic calculator on a single 'chip'. The number of components that can be combined has now reached 1,000,000 and continues to grow. Technical progress is accompanied by very rapid reduction in costs, size and energy consumption, with a simultaneous improvement in reliability (see Figures 1-2). The impact of these technologies of large-scale integration (LSI) is extremely widespread, not only in computers (Gerola and Gomory, 1984; see Table 1). The total value of production of integrated circuits by the United States reached US $27 billion in 1984.

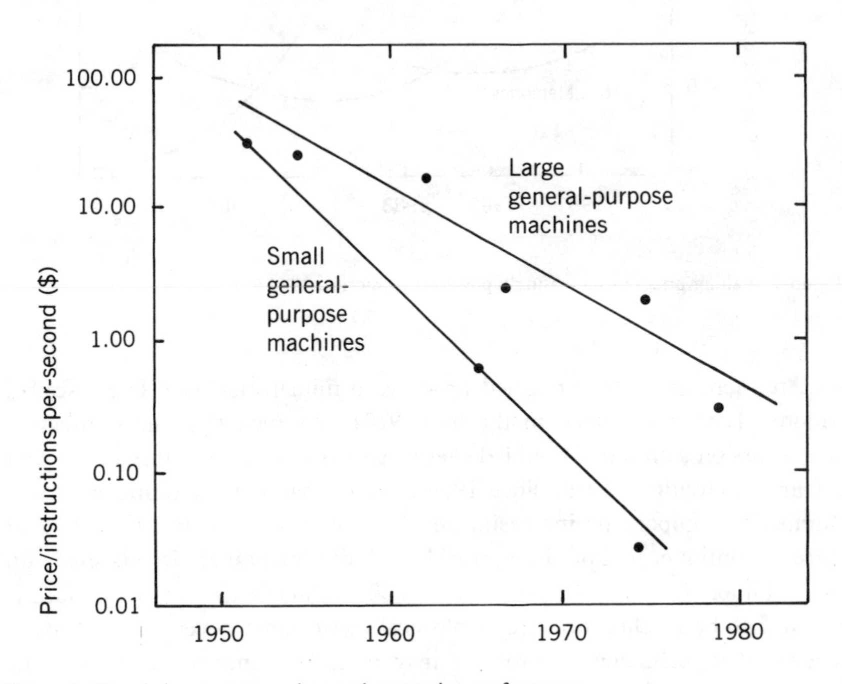

Figure 1. Trends in computer prices and processing performance.

Note :
The lower curve represents an average improvement for roughly equivalent small machines of 25 percent per year; the upper curve shows an average 15 percent per year improvement for large general-purpose machines, newer models of which offer added function as well as more power. Note that the function provided in one 'instruction' in the two cases is not equivalent.

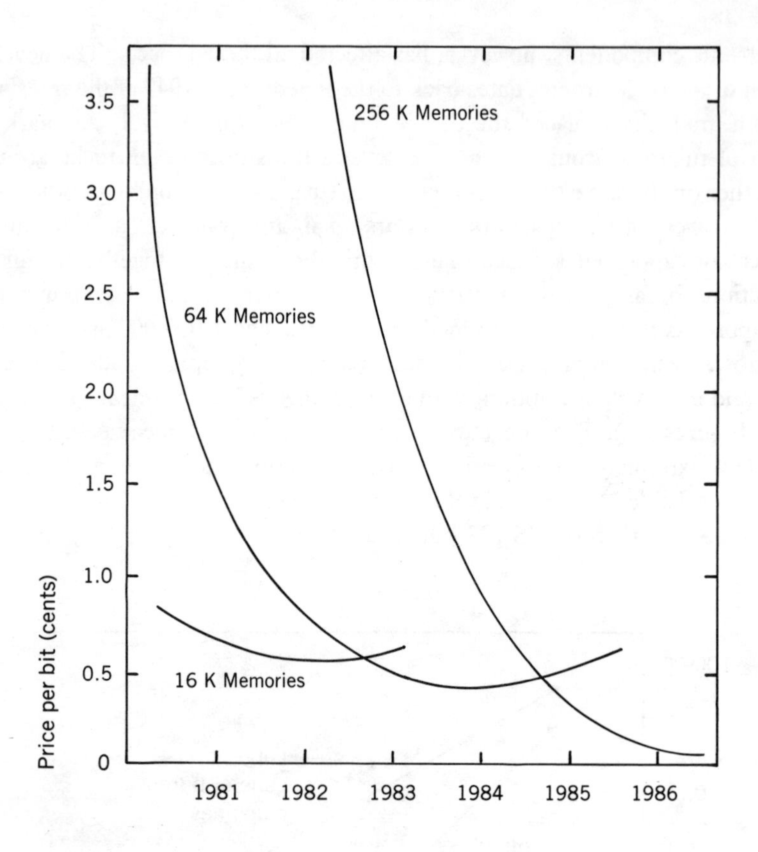

Figure 2. Falling prices of increasingly powerful memories, 1980-6.

Are there limits to continued progress in miniaturization? The so-called Moore's Law, formulated in the mid-1970s, observed that the number of transistors on a chip had doubled every year since the appearance of the first integrated circuit in 1959; since 1973 the pace has slightly diminished, the number of components increasing fourfold every three years. If this trend were to continue, it should be possible to build integrated circuits made up of 10 billion transistors by the year 2000, although this figure in fact is unlikely to be reached because of physical constraints; nevertheless, experts suggest that technological progress may permit an increase to within the region of several hundred million to a billion by the end of this century. We are thus only at the half-way stage in this major technological change.

REPRESENTATION OF INFORMATION

Information theory defines information as a physical quantity measurable in

Table 1. Microelectronic applications.

Domestic Appliances
Washing machines
Ovens and hotplates
Microwave ovens
Sewing machines
Central heating control
Gas, water and electricity meters
TV controls
Hi-fi controls
Automatic station tuning
Station indication
Alarm clock radio
Vacuum control in vacuum cleaners
Temperature and humidity control of electric irons
Door bells
Alarm systems
Lock indicators
Presence detection

Personal Use Appliances
Games of intelligence
Watches
Camera setting
Projectors
Dark-room equipment
Clocks
Hobby computers
Hobby equipment (drill, lathes, etc.)
Personal diary
Toys, talking and walking dolls

Health Care
Preventive medicine
Function testing
Patient monitoring
Cardiac monitoring
X-ray equipment (tomography)
Blood pressure monitoring
Sugar-content testing
Automatic laboratory
Warning of epileptic attack
Aids for the handicapped
Automatic prostheses
Diagnostic systems
Composition of medicines
Testing incompatibility of medicines
Therapy supervision

Education
Speaking dictionary
Translation computer
Encyclopaedia
Language instruction
Correction of spelling
Picture books
Visual training
Computer-assisted instruction/learning
Mathematics instruction
Display manipulation (graphics)
Spatial-ability training

Motor Vehicles
Routing
Anti-skid equipment
Ignition control
Fuel-supply control
Engine monitoring
Brake testing
Diagnostics
Navigation

Transportation
Traffic monitoring
Timetables
Ticket-reservation system
Departure-time information
Automatic marshalling
Air traffic control
Train control
Container control
Container freighting
Container identification
Truck identification
Truck control

Stores/Shops
Pricing
Customer/order file
Inventory
Automated accounting
Purchases and sales registration
Automatic cash register
Petrol pumps
Automatic stock control
Progressive turnover check
Reporting on buying behaviour

Telephone
Teleconferencing
Extended telephone (answering) service
Automatic calling
Automatic telephone directory
Automatic call transfer
Personal telephone number
Phone cards
Link-up with word processor
Telefacsimile
Viewdata
Electronic mail
Multi-telephone system within home and business
Intercom
Individual connection to large computer via telephone line

Safety/Security
Protection against burglary
Presence detection
Access permission
Control of access to classified documentation
Personal identification by voice and other criteria
Admission-card inspection
Emergency alerting service

Banking
Automatic teller machines
Smart cards
Electronic funds transfer systems
Banking at home
Automatic signature identification
Fraudulent money checking

Office/Printing publishing
Phototypesetting
Word processing
File control
Pre-checking of prints
Document traffic
Compilation of diaries
Electronic mail
Telefacsimile
Office at home
Electronic typewriters
Dictation equipment

Industry/Manufacture
Versatile automata (robots)
Numerically controlled machines
Weaving looms
Pattern design
Concentration regulators
Volume regulators
Pressure regulators
Air-pollution monitoring
Water-pollution monitoring

Agriculture
Feed control
Feed proportioning
Feed formulation
Cattle-condition monitoring
Crop control
Air conditioning
Weather reporting
Soil-condition reporting

Source: Adapted from *The Social Impact of Micro-Electronics: Report of the Rathenau Advisory Group*. The Hague, Government Printing Office, 1980.

the same way as energy or mass, a definition which inevitably overlooks the potential human value of the information — its significance — and which links the quantity to a purely physical process: the choice between the *n* equally probable solutions of a given problem. The unit of information, the 'bit', is the quantity of information required to make the binary choice between 0 and 1.

Léon Brillouin clearly identified the strengths and limitations of information theory:

'The methods of this theory can be successfully applied to all technical problems concerning information: coding, telecommunication, mechanical computers, etc. In all of these problems we are actually processing information or transmitting it from one place to another, and the present theory is extremely useful in setting up rules and stating exact limits for what can and cannot be done. But we are in no position to investigate the process of thought, and we cannot for the moment introduce into one theory any element involving the human value of the information. This elimination of the human element is a very serious limitation, but this is the price we have so far had to pay for being able to set up this body of scientific knowledge. The restrictions that we have introduced enable us to give a quantitative definition of information and to treat information as a physically measurable quantity. This definition cannot distinguish between information of great importance and a piece of news of no great value for the person who receives it' (Brillouin, 1956).

The quantitative definition of information thus establishes the limits of the theory's usefulness: as long as the information is circulating *within* the technical system, information theory can deal with the phenomena involved; but as soon as it interacts with human operators, psychological or sociological factors come into play which are beyond the narrow competence of the theory. Of the five categories of information transactions, three (processing, transfer and storage) are susceptible to the quantitative approach, whereas capture and retrieval lie beyond its scope whenever human operators are involved.

Analog and numerical methods

In order to register information on a physical medium, for example an electromagnetic field, the medium has to undergo a transformation and there has to be a correspondance between its physical state and its significance in terms of information, i.e. the information has to be coded in some way. This can be achieved either by the so-called analog process, i.e. by establishing a specific quantitative relationship between a physical parameter and the corresponding information (as in the case of the conventional telephone, where the information to be transmitted is a sound — a rapid fluctuation in air pressure — and the medium is an electric current: the microphone transforms the variations in pressure into variations in strength of the electric current, while the receiver performs the reverse operation), or else by symbolizing the information in the physical state of the medium (as in the case of standard scripts or the Morse telegraph system, in which each letter of the alphabet is represented by a sequence of dots and dashes without any correspondance beween the strength of the electric current and the information transmitted).

Symbolic coding was used almost exclusively until the nineteenth century, when the first analog processes of storing and transmitting sounds and images were invented (photography, telephone, phonograph). Since the 1940s the trend has been to return to symbolic coding, using the binary language in which all information can be rendered, no matter what it is, by means of the two symbols 0 and 1; the practice of converting all information into this form before processing, transfer and storage is becoming universal.

The conversion into binary coding is straightforward because many physical systems have two, and only two, states. This is the case for certain electronic circuits in particular, which have been the basis of the technology of electronic calculators. Thanks to the prominence of microelectronics, the binary system is gradually being adopted for all information technologies: long-playing records are being replaced by compact discs (on which the information is stored in digital form), and the telephone system is slowly becoming entirely digital. Television transmissions, however, are still on the analog principle, and the huge size of the existing stock of receivers will act as a brake on the evolution of digital techniques.

There is a very important socioeconomic consequence of this general trend: it means that microelectronics are the essential route to mastery of production of information systems.

Information exchange between man and machine

By contrast with traditional symbolic coding systems, binary coding is not suitable for direct use by the human brain: it can be 'deciphered' but not 'read' fluently. For information to be transmitted via a human operator, the machine code has to be converted into symbols or signals adapted to human use, i.e. in linguistic or visual form.

Language and image. Until the invention of the telephone, symbolic coding in the form of a script was required to send or record a message. Technological progress is in general moving towards the elimination of script for the transmission of information, and hence the need for recipients to be literate.

Visual images are the most significant means of transmitting information without using language (though sounds like music that are not language-based are, of course, also used); the simultaneous transmission of images is therefore a crucial advance in telecommunications. Like sound broadcasting, television does not require the recipient to be literate since it combines two sets of information, neither of them written; moreover, both messages can be recorded on cassettes. This diminution in the role of script in communication raises serious cultural and socioeconomic questions which will be addressed further below: how will this influence literacy in both developed and developing countries? what are the more general economic and cultural repercussions of these new information flows?

Qualitative and quantitative information. Another way of distinguishing the kinds of information received by the human brain depends on whether the information produces impressions that can be described but not quantified or whether it involves a measurable quantity that can be processed mathematically. The terms 'soft' and 'hard' are sometimes used to describe these two alternatives, with many gradations between them.

This distinction does not necessarily coincide with the previous one: language may convey hard information (an auditor's report, the exchanges between the control-tower and the pilot of an aircraft coming in to land) as easily as soft (a political speech or a poem); a visual image may contain soft information (a painting) or hard (a diagram). Besides, this distinction has no intrinsic significance — it depends upon the use made of the information by the brain, and for the machine it does not exist. Whether an image is to be processed or sent for artistic purposes, it still has to be coded and to undergo quantitative processing. The importance of the distinction lies rather in the degree to which the machine may be able to assist the brain according to the nature of the operations that the brain wants to perform. The machine can efficiently replace the brain if it is a matter of applying mathematical techniques to the information, indeed the computer's power and capacity are much greater than those of a human being. On the other hand, the machine is relatively inefficient when it comes to operations that the brain performs comprehensively without revealing the way it does so, as in the recognition of shapes. Nonetheless these functions of the brain can be reproduced mathematically, and the scope of computers in this area is constantly expanding. It therefore seems fruitless to pursue a precise definition of the borderline between soft and hard information.

Quantities

The use of binary coding means that the quantity of information contained in any kind of message is expressed as the number of binary symbols, or number of 'bits', necessary to code the information. Multiples of the bit (kilobit = 10^3 bits; megabit = 10^6; gigabit = 10^9) are commonly used to denominate the capacity of a computer memory or the size of a message. The 'byte', the other standard unit in current use, is the quantity of information contained in a 'word' made up of eight binary characters; one 'byte' is therefore equal to eight bits. The memory capacity of a micro-computer is normally expressed in kilobytes, often abbreviated 'k'. Generally one byte is needed to code a letter or a number.

The capacity of a telecommunications channel is defined in terms of the number of bits that it can relay in a second. Information theory demonstrates that there is a close relationship between this capacity and the width of the section of the frequency spectrum which is used for transmission, the 'bandwidth'. The frequency of a radio wave is measured in Hertz (cycles per second); the wave has to be modulated in order to carry a signal, i.e. certain of its characteristics have to be altered (phase, amplitude or frequency) depending on the signal to be transmitted. After this operation the modulated wave no longer occupies an exact frequency but rather an area of the spectrum whose width is proportional to the flux of information. The factor of proportionality depends within certain limits on the kind of modulation involved; as a first approximation, a minimum band width of 2 Hertz is required to transmit one bit per second. The relationship between the bandwidth and the channel capacity has important technical and institutional consequences.

In order to have access to the bandwidths required to send the increasing flows of information it is necessary (given the availability of the appropriate technologies) to move towards the higher frequencies in the electromagnetic spectrum — hence the gradual rise in the frequencies used for trunk transmission and broadcasting. Lasers and fibre optics, which give access to the range of optical frequencies, represent an extremely important stage in the search for higher frequencies to exploit.

Furthermore, the electromagnetic spectrum is a unique resource which should be put to optimal use for the benefit of the entire world, and hence requires coordination of its exploitation so as to avoid the interferences produced when two neighbouring bands overlap. Responsibility for this coordination lies with an agency of the United Nations, the ITU (International Telecommunications Union), which plays an important part in the development of international telecommunications.

Television broadcasting is one of the basic causes of the use of higher frequencies since a television picture contains a vast quantity of information. Take, for example, a picture in black and white made up of 600 lines, each line containing 600 points, i.e. 360,000 points in all required for the rather poorly defined image produced by a modern television set; if 8 bits (one byte) are used to define the intensity of light at each point, a message of 8 x 360,000 = 2,880,000 bits is necessary to transmit one picture, so that at 25 images per second the flow of information works out at 72 million bits per second, whereas a telephone signal requires only 50.10^3 bits per second.

Now that we have defined the units of measurement for the contents of a message or the output of a communications line, the more difficult task remains of describing the capacity of computers in quantitative terms. The powerfulness of a computer cannot be completely rendered by a single figure, though several different units have been used to measure the overall speed of processing; the number of basic operations per second that the machine can perform or the number of 'flops' (floating point operations) is often used. Modern super-computers now reach processing speeds in the tens of millions of flops (60 million flops for Control Data's Cyber 205, 20 million for the Cray 1), and in the 1990s it is hoped to attain speeds in the tens of gigaflops (10^9 flops). Such speeds will need to be matched by corresponding speeds of capture and retrieval, while the connection of machines sited in different geographical locations presupposes the availability of very large capacity telecommunications networks.

INFORMATION TRANSFER TECHNOLOGIES

The logic of developments in telecommunications follows obviously from the physical properties of electromagnetic signals. Under normal conditions the signal tends to move in a straight line; in order to use it to send a piece of information between two distant points on Earth, one must either guide it along the Earth's surface or else connect the two points with a system of relay stations visible to each other. Within a certain frequency area which corresponds to the kilometric to decametric waves, there is a natural wave guide formed by the two conducting surfaces — the ground and the ionosphere. These frequencies are suitable for transmitting speech, but for larger quantities of information (as for television pictures, for instance) higher frequencies must be used, and at these higher frequencies the ionosphere is transparent and the signals disappear into space. Furthermore, the importance of diffraction phenomena, which weakens the shadow effect of variations in relief, diminishes as the frequency rises. In sum, for metric frequencies and above, the range of a transmitter is limited to the area within direct view of the antenna.

Obviously the range can be extended by using high points on the Earth's surface, as was done at the end of the 18th century for Chappe's optical telegraph, but the limits of this technique are quickly reached, besides which the relief creates hidden pockets as well as high points. In order to reach an area as large as France (550,000 km^2) with a national television coverage that still has many blindspots requires about 100 transmitters and 1,000 automatic relays. In addition, a network of radio links is necessary to send the signals between the studios and the network of transmitters.

A satellite, by contrast, serves its whole area of operation at a single stroke;

within that area, telecommunications stations may be established as desired, and the cost of the system is a function of the number of stations but not, in the first instance, of the distances between them. The basic rigidity of the system derives from the fact that the element in space (the satellite(s)) has to be put in position once and for all even though the level of use may be very low as long as the terrestrial element (the network of ground stations) is not fully deployed. To take the example of television satellite broadcasting: individual users will not begin to equip themselves until the satellite is in orbit, so that for an initial period of variable duration, a very costly investment will benefit a limited number of users.

The nature of satellite telecommunications obviously has much in common with air transport: both make it possible to avoid the construction of continuous infrastructure facilities on the ground. Use of satellites therefore permits developing countries to acquire at once modern telecommunications systems — whatever the inherent problems of their terrain.

The borderline between space and terrestrial telecommunications is very fluid since it depends on technical progress as well as their inherent characteristics.

Fibre optics

Use of optical frequencies is the most significant development in land-based techniques. The technique is based on two discoveries: lasers and optical fibres. Lasers involve the same basic technologies as microelectronics; they are used to insert the signals for transmission into the optical fibre, and optic transducers pick up the signal at the destination and translate them into electronic signals.

Optical fibre made its appearance in the early 1970s when the Corning company managed to produce glass that was sufficiently transparent to allow signals to be sent over great distances. The most important characteristics are:

— the vast capacity for transmission linked to extremely high frequencies; it may be possible to reach 100 million bits per second on a single fibre

— the minimal weakening of the signal over distance, which means that the number of repeaters can be reduced, from one every 2 km on a coaxial cable to one every 40 km on the optical fibres already in commercial use; distances of 100 km have already been reached in the laboratory

— the use in very small quantities of abundantly available materials for the production of the fibres and associated equipment, whereas the traditional telephone lines, coaxial cables, require large amounts of a rare metal, copper.

— the immunity from environmental interference.

The conjunction of these remarkable features suggests that optical fibres may become the universal medium for sending terrestrial information.

Progress in this field can be appreciated by considering two events: (1) The first transatlantic telephone cable (TAT1) was put into service in 1956 between Newfoundland and Scotland; it could carry 36 simultaneous telephone conversations but could not transmit a television picture. Until 1964, and the launching of the Intelsat system of satellites, there was no means of transmitting a television programme or having telephone contact across the Atlantic except at a low quality level by short-wave.
(2) The first transatlantic cable using fibre optics (TAT8) will be put into operation in 1988; it is designed to carry 40,000 telephone circuits at a cost that is less than half that of its predecessor.

Telecommunications satellites

So far space-based communications technologies have been used almost exclusively for information purposes. 'Application satellites' fall into two categories, depending on the kind of signal that they receive: either 'natural' signals (i.e. not deliberately beamed at the satellite) on the basis of which information is then transmitted back to Earth about the physical environment or certain kinds of human activity; or else signals deliberately beamed to the satellite for retransmission. Observation satellites (meteorological, remote sensing, military observation) belong in the first group; telecommunications satellites in the second.

Almost all of these telecommunications satellites make use of the geostationary orbit, which is such that thay appear stationary in the sky to an observer on Earth. This circular orbit, at an altitude of 36,000 km at the level of the Equator, is unique, so that — like the electromagnetic spectrum — it represents a limited natural resource that should be put to optimal use. The ITU is also responsible for the coordination necessary for the distribution of places in the geostationary orbit and frequencies for telecommunications satellites.

Progress in the area of space telecommunications arises from the conjunction of two factors: the improvement in the efficiency of satellites (growth in their size and in the electric power available brought about by the greater capacity of the launchers; growth in the complexity of the payload brought about by technical advances) and the concurrent developments in terrestrial technologies.

The limited capacity of the first satellites, at the beginning of the 1960s, required the use of high performance ground antennae, of immense size (32m in diameter) and at vast cost. Their use was thus feasible only for intercontinental communications, where there was no alternative; they

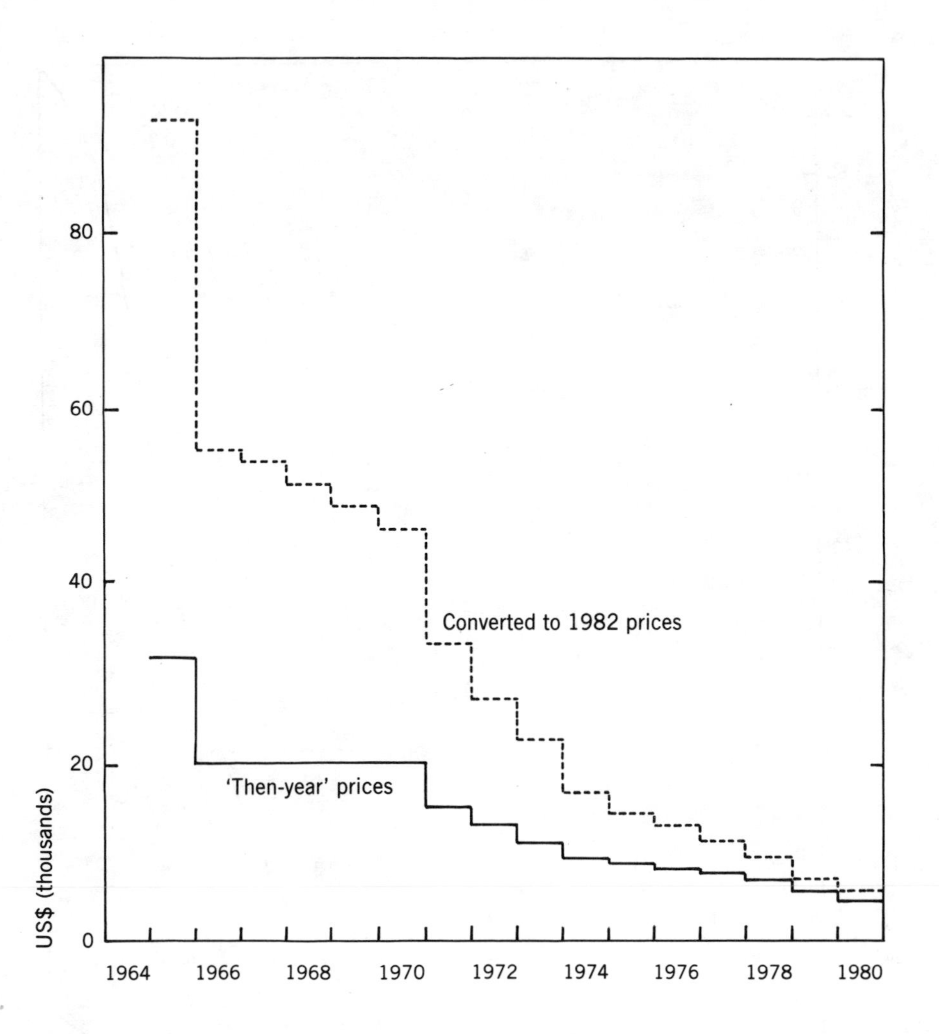

Figure 3. Intelsat annual charge per unit of utilization for full-time service expressed in current prices and in real terms.

nevertheless developed rapidly with the Intelsat system from 1964 onwards (Figures 3-4).

Advances in space technology have made it possible gradually to increase the power supplies on board the satellites. The potential has been exploited in two ways: to expand the capacity of the intercontinental telecommunications system (Fig. 4) and at the same time reduce the specifications of the ground stations (Intelsat now uses antennae of 15 m diameter as well as 32 m); and to extend into new areas of use which require less massive and costly ground installations, thus giving to the satellite an increasing share of the efficiency of the system. As a result, following about a decade after

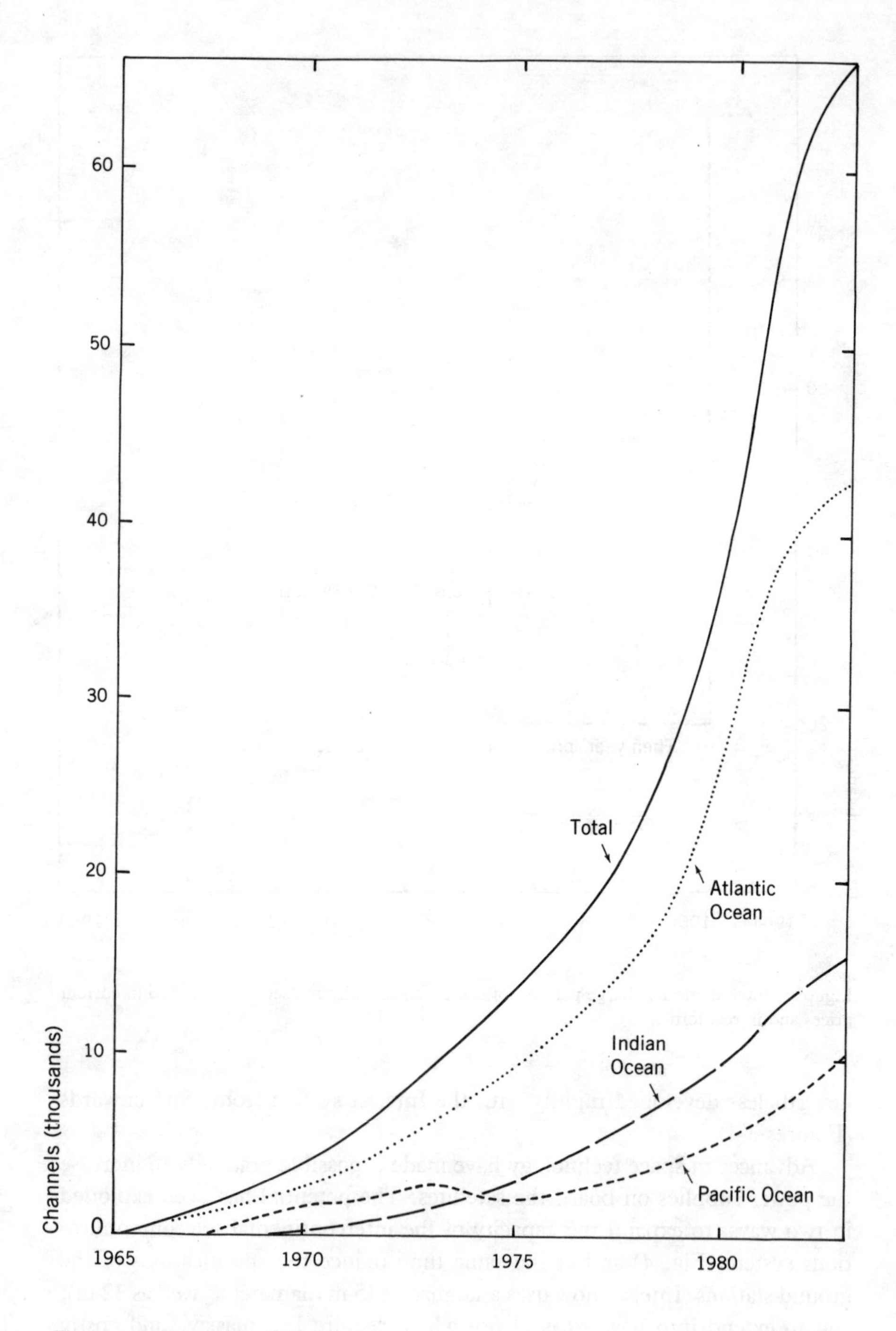

Figure 4. Intelsat traffic, 1965-83.

intercontinental telecommunications, national (so-called 'domestic') systems have come into being, serving countries whose extent or whose terrain make land-based systems difficult (Canada, Indonesia), and then serving the most highly developed nations (the United States of America, Europe with the regional system Eutelsat, France with Telecom 1). The antennae needed for these domestic systems are only about 2 m in diameter, which facilitates their rapid installation and multiplication. Space links with ships at sea have been developed simultaneously with the intercontinental Inmarsat system.

Ultimately individual users should be able to receive the television signals, but for that the antennae would have to be much smaller (0.6-1 m diameter) and the receiver less expensive, perhaps cheaper than a colour television set. The satellite would have to transmit a strong signal via directional antennae that focussed the programme on the zone to be served.

The essential difference between the domestic telecommunications satellite and the broadcasting satellite is that a two-way link is possible with the former, allowing interactive usage, whereas the latter is limited to broadcasting from the centre outwards and cannot feed back the recipients' reactions to the message.

The growth in quantitative efficiency of the telecommunications satellites proper (both domestic and intercontinental) is matched by increasing complexity of performance: the first satellites of the Intelsat system were in effect equivalent to a 'cable in the sky' linking two stations on either side of the ocean; present-day satellites are real central switching stations that allow any two stations in the network served to link up through multiple access facilities.

Current trends

It is extremely difficult to foresee how space and terrestrial technologies are likely to develop in the provision of future telecommunications systems. From a purely technical angle, the problem is dominated by the fact that the two technologies are competing with each other and both are still evolving rapidly — both fibre optics and telecommunications satellites. Furthermore, powerful and diverse political repercussions cause governments and administrators to become involved in the decisions to be taken. The choice between direct satellite television broadcasting and an optical fibre system is the archetypal example of a decision where political and cultural considerations, and lobby groups, carry more weight than technical and economic reasons.

In any event, the combination of terrestrial and space technologies together permit virtually limitless growth of overall information flows: even an increase by a factor of 1,000 or 10,000 over current levels would not encounter any serious physical barrier.

INFORMATION PROCESSING TECHNOLOGIES

The computer was invented when John von Neumann conceived the idea of putting the instructions to run an automated system in electronic form. It is the use of a 'program' (recorded, easily altered, repeatable) that is the fundamental difference between computers and earlier forms of calculator in which the information was stored in a mechanical form; we see again here the phenomenon of 'dematerialization' of information which characterizes all current technical progress.

Computer technologies are organized round this basic distinction between the program on the one hand and the machine that uses it on the other, between the 'software' and the 'hardware'. Overall progress requires both elements to develop; in spite of their interdependence, each is derived from a separate logic.

Hardware

Developments are mostly determined by progress in electronic technologies. With regard to the technology of the circuits that carry out the processing tasks, all of them using a binary language, four generations of computers are traditionally identified. The first used vacuum valves (the last representative was the IBM 704 in 1956); the second was based on the transistor — Univac produced the first model in 1959, the year that saw the first integrated circuit, which in turn was to be the basis of the third generation. The launching of the first microprocessor by Intel in 1971 marked the start of the fourth generation, in which progress has been governed by advances in large-scale integration technologies and the development of silicon chips that unite the processing and memory functions. Two other equally important aspects of computer history are not brought out by this approach that concentrates on the progressive miniaturization of the electronic components: the increase in processing speeds, and the expansion of memory capacities and quicker access to the information stored in the memories.

Computers are usually divided into five major categories according to their power and price: pocket calculators, micro-computers, mini-computers, main-frames and super-computers. The price differential between the two ends of the range is equivalent to that between a bicycle and a jumbo-jet.

The main-frames are the big machines capable of handling the demands of several users simultaneously (multiple processing), fed in via multiple access; they constitute a substantial share of the market and represent the culmination of the general tendency to expand processing capacities. The super-computers, which account for a tiny fraction of the market (only 130 exist worldwide in 1985), are designed to reach the maximum possible processing speeds, currently at the expense of flexibility. They are intended to cope with problems of simulation that require vast quantities of complex data to be processed all at once, for instance simulating atmospheric movements for weather-forecasting purposes, or in wind tunnels (for simulation of air flows round the body of an aircraft). In this regard super-computers have important implications for both industry and national defence in the future.

At the other end of the range, the micro-computers make use of the miniaturization of electronics built round a microprocessor; they are highly transportable, cheap and within the reach of individuals or small businesses, though they are also capable of considerable computing tasks. The market is growing rapidly and is likely to outstrip all the other categories; at the same time, advances in microprocessors are expanding the capacities of this type of machine. Mini-computers occupy an ill-defined zone between the main-frames and the micros.

Two opposite trends characterize the evolution in hardware: one, apparent from the outset, towards concentration and large size embodied in the main-frame; the other more recent, towards decentralization and accommodation of specific requirements expressed in the growth of micro-computers. Equilibrium between these two trends has not yet been achieved, and it is difficult to foresee the ultimate outcome.

Software

Developments in software are intimately bound up with advances in hardware, but they have no direct links with the electronic technologies that have stimulated changes in hardware. Software is derived from mathematics, especially the branch of mathematics concerned with expressing logical operations in symbolic terms in the manner of Boolean algebra.

Each kind of computer comes with its own set of programs that regulate the use of the hardware; the creation, implementation and use of this operating system are inseparable from the hardware. An end-user program has to be loaded into the computer's memory before it can carry out a given task. The computer responds to instructions coded in binary that are extremely laborious and difficult for a human operator to use, so that languages had to be developed in order to allow non-specialists to program the machines; these had to be both easily understood by the operator and easily translated into machine language by a special program, the compiler. The creation of programming languages is an activity at the edge of research and industry, where universities have played a very important role.

The creation of programs designed to deal with specific problems has become the basis for an important branch of economic activity. The number of computers is growing much faster than the number of users able to program them, with the result that there is scope for an industry to produce programs. Two sorts of enterprise have evolved: firms that create programs to the order of individual users, usually dealing with medium-sized or large computers and with firms that have computer equipment but not the specialists necessary to produce the programs that they need; and firms that create packages of standard programs for general sale. The need for packages has developed from the spread of micro-computers — the cost of 'made-to-measure' programs is beyond the means of the owners of micro-computers, hence the development of ready-made packages whose high costs can be spread over a wide market. The market for packages has grown very rapidly, roughly 50% per annum with an estimated turnover of US $6.5 billion in 1988.

The availability of varied and efficient program packages is as important a determinant of the commercial value of a micro-computer as the technical quality of the hardware.

In the commercial battle between the manufacturers of hardware, the choice of standards, and hence the adoption of a strategy of compatibility or incompatibility with competitors' products, has been a very important aspect in the development of computer and microelectronics and one which has eluded state efforts at control and standardization. It is only when the standards affect telecommunications that public authorities have been able to intervene. The result for users, as many developing countries have learned from bitter experience, is the risk of premature obsolescence of their chosen equipment when the particular standard is abandoned following a purely commercial defeat, as well as problems of incompatible spare parts and accessories ordered from abroad. It seems likely that this situation will gradually improve in time, though this provides little comfort for those who have already invested in costly systems based on earlier norms.

Interface between man and machine

Communication between man and machine has evolved with technical progress, combining developments in both software and hardware. The main stages of development have been:

— languages that more closely resemble human languages

— non-linguistic forms of communication, in a similar vein to developments in telecommunications. These include methods of both input (light pens which can be used to draw or write directly on the screen, screens that react to finger pressure, the 'mouse' that translates movements of the hand on a flat surface into movements of a cursor on a screen) and output (voice synthesis, use of graphic symbols that are immediately recognizable). All of this is based on relatively simple technologies and can be used with a micro-computer.

Voice command of machines poses much more complex problems but the technique is emerging nonetheless.

The range of trends grouped under the general heading 'user-friendly' (by contrast with the forbidding relations based on the keyboard and the use of peculiar languages), together with the use of program packages, is bringing the computer within the reach of everyone, even without proper training, and soon even of the illiterate.

Another development is the attempt to simulate human intelligence on the computer, and to extend the competence of the machine to handling problems that cannot be put in straight mathematical terms — the field of 'artificial intelligence'. An early example is the 'expert systems', which first record the knowledge of human experts on a given subject and then simulate their reaction to a particular situation. Comprehension of speech and recognition of shapes are also part of this field, which is still in its early stages. It involves the gradual extension of the area of competence of the computer, moving from mathematical calculations for which the brain is not particularly efficient towards areas in which the brain excels and which can be described as judgement, common sense, the knowledge of experience. This trend, linked with developments in user-friendly techniques, means that more and more is available to the user, while less and less is required of him.

Information exchange between machines

The appearance of the micro-computer based on the microprocessor has generated a dispute that is almost ideological between the partisans of centralized computer use based on main-frame machines linked to terminals and the partisans of decentralized use, with dispersed independent processing units. The choice between these two extreme alternatives highlights certain repercussions on social organization, and elicits preferences that are vigorously expressed (Lussato, 1983); there are implications, too, for some branches of administration, for example those that control communications networks. The dispute is now dying down, and a certain balance has been achieved among the territories of the different sizes of machines. Besides, micro-computers rapidly acquired the possibility of being linked into a network so that the distinction could no longer be made so easily between the terminal of the main-frame and the completely independent small machine. There remains an overall trend towards interconnected machines that creates demands on the telecommunications networks, whether it involves giving micro-computers access to large central data-banks or else dividing the work load among large machines that are geographically dispersed, or perhaps grouping the machines to make up a single system with permanent links (as could happen with some military systems). All in all, we are obviously just beginning to see the potentially vast scale of need for information transfer that is emerging.

APPLIED TECHNOLOGY: ROBOTICS

The element that distinguishes modern robots from the automated machine tools of the past is the fact that they are linked to computer technologies and can be reprogrammed to perform varying tasks; they can also communicate directly with other computerized devices (OTA, 1985; Draper, 1985). Most robots currently consist of one or more articulated arms attached to a base (the manipulator), which are governed by a computerized controller and powered by electric motors, pneumatic or hydraulic systems. A 'gripper' or 'hand' can wield tools or move parts. The so-called 'second generation' robots have some additional capacity to sense shapes so that they can assemble quite complex items (for example, 90% of a personal computer can be put together by a robot, and in Japan, robots are already being used to make other robots).

The potential applications are virtually limitless, though for the moment the main use has been in repetitive precision tasks like arc- and spot-welding and dangerous or unpleasant operations such as moving molten steel pieces through production processes or paint spraying. The car industry was the quickest to adopt the first-generation robots, and consequently machines have tended to be designed to meet that industry's particular needs. The more sophisticated robots with sensing devices are increasingly used in light-manufacturing, especially microelectronics, and even in agriculture for harvesting and irrigating. Obviously, the great advantage of robots is that they can work continuously without loss of concentration, fatigue or disease, in conditions that human workers could not tolerate. The practical limitations include the inability to lift weights greater than approximately 2kg, and even advanced robots still have a quite limited capacity to 'see' in comparison with human beings, though improvements are being made all the time. Work is underway on developing machines that can respond to simple spoken commands, which would allow direct control by operators and eliminate the need for specialized programming. Mobile robots that can fetch and carry, or clean surfaces, are also being developed.

The major impact of reprogrammable robots is on relatively small-scale 'batch' production where their flexibility permits production lines to be switched very easily, hence they are ideal for the increasingly customized products characteristic of developed economies. Such adaptability is not required for mass production, where it makes greater economic sense to instal 'hard' automation, i.e. specialized machines capable of a single task.

Clearly these sophisticated machine tools can confer enormous benefits if used independently for certain tasks within a conventional production process, but they are most advantageous when combined with other computer-based operations which cover the whole spectrum from design to final product. It is possible to sketch a design directly onto a cathode-ray tube linked to a computer, which turns the sketch into an electronic blueprint that can be revised endlessly as different shapes and ideas are tried out. The computer can then produce a program directly from the drawing in order to set up the machine to make the part, or to coordinate a series of robots to make and assemble several parts. Management can meanwhile monitor output, inventories and sales from information that is held centrally on the computer and constantly updated, instead of being recorded separately at every stage of production and marketing. There are as yet very few examples of fully integrated systems since the technical problems of coordinating their very complex interactions have not been entirely satisfactorily solved, but the basic principles are clear.

Mastery of technical change

CRITICAL TECHNICAL DEVELOPMENTS

As we have seen, from a technical point of view the critical factors in the remarkable progress of information technologies have been the miniaturization of electronics and the combination of computers with telecommunications. Miniaturization has led to phenomenal improvements in the efficiency of machines, the development of small computers and a generally much wider range of applications of great versatility, all at a lower cost which puts the equipment within reach of a much larger public. The link between computers and telecommunications promises to transform the speed and scale of information transfer. Certain other trends are emerging: a tendency towards 'dematerialization', as use of a tangible medium is abandoned across the whole range of information transactions; reduction in the need for coding, or indeed language, in the communications between human beings and technical systems; universal use of binary programming; continuing extension of the capacities of information processing systems as they move away from strictly mathematical functions to replicate the higher aspects of human intelligence. Research on the 'fifth generation' of computers, on artificial intelligence and on expert systems may give new impetus and dimensions to the already pervasive impacts of the information revolution.

This supply of new technologies ('technology push') cannot, however, be separated from the social demands ('market pull') that have contributed to their growth: initially from the military, then from large organizations (both public and private) stimulated by the growing complexity of their management needs, at both the national and international level.

In most of the industrialized countries the net effect is that the total share of information-based jobs has grown very considerably as a result of both the expansion of employment in the service sector and also the replacement in manufacturing industry of 'on-informational' by 'informational' labour. Today one out of every two American workers — as against one in eight in 1900 and one in three in 1950 — produces, processes or transmits information. The proportions are roughly the same if one takes instead the share of value added by the information and communications sectors in total GNP (Figure 5).

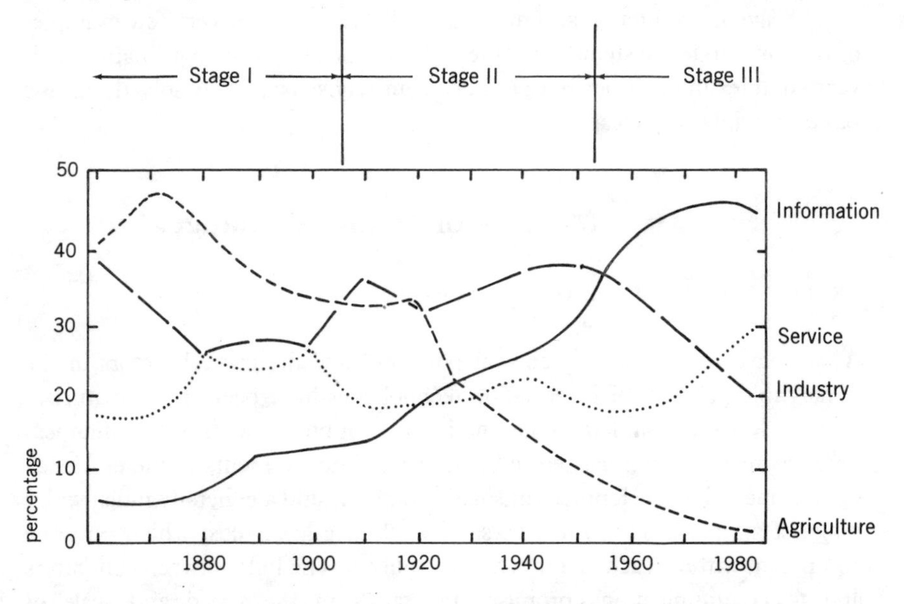

Figure 5. — Four sector aggregation of the American work force by percent (using median estimates of information workers).

In this sense, whether we call it the 'informatization' of society, the 'intelligence revolution' or the shift to the post-industrial society, the transformations that result from the progress of information and communications technologies are not in essence different from those that have propelled the increasing mechanization of labour since the start of the Industrial Revolution. Nor is the interaction new between technical change and economic and social development: the changes that arise from automation are accompanied today in the field of information — as were earlier changes in the fields of energy and matter — by social and cultural adjustments of such magnitude that they must be seen as both cause and effect of these technical and economic transformations.

MASTERY OF PRODUCTION AND MASTERY OF USE

The remarks made in the introduction about the prerequisites of modern high technology research and production are especially true in the information technologies. Computers and telecommunications in particular require a large pool of highly qualified scientific and technical labour and also institutional structures (fiscal, financial, banking) disposed to take on very high risks. Product lives are short, with the result that firms have to be capable of maintaining their innovative edge and at the same time of distributing their products rapidly worldwide. The pace is daunting for many industrialized countries, and all the more so for most developing countries — indeed, the socioeconomic transformations that the information technologies are generating are such that any industrialized country that cannot gain mastery of them will end up in a state of under-development. Even among the industrialized countries it is not certain that every one will be able to make the requisite changes to their systems of production in order to accommodate the production of the new goods and services created by the information revolution.

In outline, two elements may be distinguished in the transformation of manufacturing and services resulting from computer and telecommunications technologies: one concerned with the application of these technologies to various branches of economic activity, the other concerned with the actual production of these technologies — in the design, creation, development and marketing of new computer products whether 'hardware' (which has material substance and value) or 'software' (the value of which resides in intellectual and not tangible properties).

Two categories of mastery correspond to these two elements, the mastery of *use* (that is, the ability to integrate the imported information technologies into the economic system so as to increase its productivity and competitiveness) and the mastery of *production* (that is, the capacity to adapt the system of production so as to produce computer-based goods and services). This distinction between the stakes involved in the mastery of use and of production is neither new nor peculiar to information technologies: in transportation, for instance, mastery of air transport is widespread while the production of civil aircraft, especially certain types of aircraft such as long-haul carriers, is highly concentrated.

Because the information technologies have repercussions throughout the socioeconomic system, far beyond the scope of earlier technical changes in manufacturing technologies, mastery of the use of these technologies is indispensable henceforth for all kinds of economic development: it would be impossible to embark upon a national-level economic policy that failed to take account of this fact, and equally impossible for the planning and management of a medium-sized firm. The possibility of modernizing agriculture, the process of industrialization itself, the efforts to manage the use of natural resources more efficiently, as also the strategies based on the improvement of traditional technologies — all of necessity pass by way of the information technologies. As long as the aim is to make the economic system more productive and competitive, there is no alternative.

Mastery of production, on the other hand, cannot become widespread. The majority of the industrialized countries do not themselves possess more than a fraction, often modest and highly specialized, of what is needed to produce the computer-related materials that their economies require. This situation is even more apparent for developing countries. In fact, only the newly industrializing countries are in any way capable of sharing in the production of computer-based goods and services, whether to their own designs or copies. Furthermore, these disparities between the capacity to produce and to use can be expected to widen, even between the industrialized countries themselves.

The general tendency towards concentration of capacity in design and production promises to become particularly marked in the case of information technologies. Throughout the whole process from R&D to production there are numerous possibilities of feedback which tend to widen the gap between the countries in the lead and those in the rear. For instance, the LSI and VLSI ((very) large-scale integration) circuits are so complex that they cannot be created without large computers, and in turn the design of super-computers requires the production of special circuitry. The industrial firms that manage to master both aspects of this process therefore possess a substantial advantage over those that have mastered only one — this is one of the key factors in the lead that Japanese industry has gained in the area of super-computers.

The creation of program-packages might appear to be more of a craft activity where small firms could be encouraged to imitate the example of Silicon Valley with its proliferation of small businesses. However, as already mentioned, it is also possible to produce packages by setting up factories for the purpose, using teams and libraries of re-usable programs in conjunction with a suitably modified telecommunications system — productivity then grows by several orders of magnitude. Even the production of hardware can be adapted to highly sophisticated robotics, through the computerization of the production process. To master this aspect of information technologies — and even more so the production of the new technical systems required for telecommunications networks — is clearly out of the question for many countries because of the skills, the organization and the financial investment involved.

While it is true that self-generating effects of this kind can also be identified among the stages of technical change in the past, particularly since the Industrial Revolution, these effects have never been so intense or so systematic: the very nature of the new technologies thus contributes directly to making the international division of labour more differentiated than ever as certain technical innovations are limited to a few countries and even to a few firms. The increasingly scientific character of the process of industrialization reduces even further the links between the mastery of use and of production of the new technologies. Just as there are fears that a 'dual society' may develop in the industrialized countries — in which the disparity in knowledge, position and status of professionals as against the rest replaces the inequalities based on birth — there could be a 'dual geography', in which only a handful of countries are able to produce high-technology products, while the rest are out of the running, even if they are industrialized, except for competition in the mastery of use.

Situation of the developing countries

Information and communications technologies have been created and developed in societies whose features are the exact opposite of those in the developing world: a high level of technological potential linked to extremely diverse academic and industrial structures; a relative scarcity of labour; considerable capital resources; an economy in which the service sector now offers more employment than the manufacturing sector or, by far, agriculture (*Futuribles*, 1984). In the United States of America, for example, the proportion of the active population engaged in services is now roughly 70%, i.e. the same proportion as was employed in agriculture a century ago (only 30% of the population produces actual goods — mainly manufactured products — and barely 3% are engaged in agriculture). The experience has been comparable elsewhere in the developed world.

In most of the developing countries, on the other hand, the structure of the labour force is very different; and it is indeed possible to define underdevelopment as, among other things, the absence of those very characteristics that have permitted the expansion of computers and telecommunications in the industrialized countries. For societies where the majority of the population remains concentrated in the agricultural sector, where the academic and industrial structures are perforce unsophisticated, where capital is limited and or labour abundant, the introduction of computer-aided manufacturing and the mechanization of labour in the fields of information and communications could have results that were more negative than positive.

For the developing world, the same polarized attitudes to the potential effects of the information revolution technologies are to be found as have emerged in the developed nations: the one optimistically sees information technologies as the one great cure-all for the problems of development (Servan-Schreiber,1981); the other pessimistically sees the new technologies as the means of reinforcing the disparities, the political and economic dependence, and the increasing loss of cultural identity of the developing countries (Mattelart, 1977). The worst is not always inevitable, it is true, and there is no reason to expect the technologies alone — whatever they are — to cause fundamental changes to the power structures governing relations between nations any more than to overthrow the organization of relations between different social groups and between individuals within a country.

Technology transfer is nonetheless a particularly delicate matter with regard to the information technologies, because of the strongly dominant position of the North and because of the peculiar nature of information which together make it especially difficult to consider the associated technologies as neutral: since it concerns information and communications, know-how and intelligence matters, access to these technologies is synonymous with power or various aspects of power — economic, political, cultural, religious, military. From the computer to the video-recorder, from television networks to telecommunications satellites, will control over the medium mean control over the message? Even the developed countries are concerned about these matters, so that it is not surprising that less developed countries feel extremely vulnerable.

DISPARITIES OF ACCESS TO THE INFORMATION SOCIETY

There are three basic prerequisites for full participation in the information society: access to a telephone, a television set and a computer. In the provision of these fundamental items the developing world generally lags far behind most of the industrialized nations.

Telephones

A telephone is indispensable because it is via telephone links that it will be possible to have access to data transmissions, interactive communications, videotex, electronic mail, data banks and the like, on both a national and an international level. In 1981, the developed countries had an average of 46 telephones per hundred inhabitants (in the United States of America 92% of households have a telephone), whereas the average of the less developed countries was 2.8. The telephone density in Africa was 0.8 per hundred inhabitants, 2.0 in Asia (excluding Japan and Israel), and 5.5 in Latin America. In addition, a high proportion of the telephones in these countries are concentrated in large urban areas: in Mexico, for instance, there are still more than 30,000 communities with no access to a telephone at all (Saunders et al., 1983). Thus the vast majority of the populations of the developing world is excluded by definition from the benefits of the information society (Melody, 1983).

Television

In 1982 the world's 600 million television sets were very unevenly distributed: 90% of the receivers were concentrated in 15% of the countries, while the other 85% (the less developed nations) owned the remaining 10% of sets. In 1980 one person out of 500 in the developing world had a television as against one out of two in the developed nations (Unesco, 1982).

A television screen gives access not simply to the broadcasts of television programmes (via conventional means, cable or satellite), but also to use of computers, to a whole range of data flows in combination with the telephone system, to video-cassettes and video-games that provide an enormous variety of possibilities for education and entertainment. The potential cultural consequences are clearly immense.

Computers

The distribution of computers is even more skewed towards the developed world, especially the United States of America — which had 60% of all the machines in 1980 while the developing countries together had only 4%. In terms of systems per million inhabitants, the United States has 248 as against 15 per million in the best equipped region of the Third World, Latin America. Furthermore, the majority of the systems in the developing countries are owned by the state and are used purely for administrative data-processing purposes (*Futuribles*, 1984); the ownership and the use are much more diverse in the industrialized countries. The possible applications are limitless — with the concomitant that exclusion from these facilities can only be increasingly detrimental to development.

Clearly simply to obtain the basic equipment for better mastery of use will require a major effort on the part of developing nations. In this they will be handicapped by other disparities, in income and education. The developing world's share of global expenditure on science and technology is about 3%, with only 13% of all the scientists and engineers. Besides, a high proportion of this tiny share is accounted for by only four or five developing countries (such as India, Brazil, Mexico) (Rada, 1983). Given limited resources and large debt burdens, the developing countries are not in a position to invest

Table 2. Date and make of earliest computer installations in selected developing countries.

Country	Private sector	Public sector	Para-public sector
Philippines	1950 IBM 650		1962 IBM 1401
Islamic Republic of Iran		1954	
India			1955 Institute of Statistics, Calcutta 1959-1962 ICL
Morocco		1957 IBM	
Mexico			1958 IBM
Chile			before 1960 IBM 1967 IBM 360
Tunisia	1960 BULL	1962	
Venezuela			1960 IBM
Argentina	IBM		1960
Panama			1960 IBM Tax authorities
Malaysia			1961 IBM 1440 National Electricity 1970 2 IBM, 1 ICL 1904
Algeria		before 1962 BULL	
Kenya	1964 East Africa Power Co.	1962 ICL East Africa Ports and Railways	
Nigeria			1962 ICL
Ivory Coast	1962 BULL Gamma 30		1962 Ministry of Finance
Singapore			1963 Ministry of Finance
Ghana		1963 ICL	
Jamaica	1963 IBM West Indies Sugar Co.		1963 IBM University of West Indies
Tanzania		1965 ICL	
Iraq		1966	
Burma			1970 ICL Ministry of Education

Source: Futuribles, 1984.

heavily in R&D, or in telecommunications systems and computer networks which — in spite of cost reductions and technical advances that eliminate some of the practical difficulties — are extremely expensive to instal and maintain. These disparities at the national level are even more dramatic at the individual level: for instance in Mexico the minimum annual wage is US $1,300 and the cheapest computer costs $2,500.

Reference has already been made to the high level of education required in order to operate and maintain, let alone design, modern technologies; this should be set against the fact that in a relatively advanced part of the developing world, Latin America, the average level of schooling is three years during which a basic mastery of reading, writing and arithmetic is achieved — hardly an adequate basis for using a computer or repairing a television set.

GROWTH OF THE INFORMATION SECTOR

In spite of the obstacles and the handicaps, some parts of the developing world have acquired varying degrees of mastery of both production and use of the information technologies proper (not, therefore, the techniques related to advanced telecommunications systems, space exploration or sophisticated armaments systems). And this is not a recent phenomenon — it dates back to the 1960s (Table 2).

The computing capacity of some Latin American countries was in fact built up at the same time as that of the most advanced industrialized countries. The absence of local skills and the complexity of the new industrial sector explain why at first all the necessary equipment had to be imported — indeed it was not until the new electronic systems were standardized and miniaturized in the 1970s that certain countries began to consider establishing their own computer industries.

Aside from the subsidiaries of the multinational corporations, the main computer user from the moment of their introduction was the State, from the administrative services to nationalized industry. In most cases, major data-processing centres were established at the same time that the first big machines were imported, thus concentrating virtually all the equipment and trained personnel in one place, a tendency that clearly arose from the nature of the cumbersome main-frame machines of the 1960s. Initially the developing world's stock of big all-purpose main-frame computers was almost entirely North American (90%), IBM alone accounting for 63.3% (*Futuribles*, 1984). The remainder were either British (ICL) or French (CII), supplying their former colonies. The situation regarding software was almost identical.

In the first decade of computer expansion, by far the best equipped region of the developing world was Latin America (58%); Asia accounted for 28%, the Middle East 8.14%, Africa 5.37%. Moreover, a very few countries in each of these regions possessed the bulk of this stock: in Latin America the four leading countries in descending order of importance were Brazil, Mexico, Venezuela and Argentina; in the Middle East, The Islamic Republic of Iran, Egypt and Turkey owned 50% of the region's total stock; in Africa, Algeria, Nigeria and Zambia made up 54% of the market. In Asia, computers were more evenly distributed among countries (Table 3). The annual growth rate of this stock over the last two decades has been very rapid (25 to 35%).

As just mentioned, information technology has been adopted mostly in public administration and in large firms (public or private) with access to international markets. The applications have therefore been largely confined to the needs of management and basic administration (payrolls, accounting,

Table 3. Distribution of large computers.

Region / country	Number
Latin America	
Brazil	1,569
Mexico	526
Venezuela	289
Argentina	194
Colombia	121
Chile	117
Peru	61
Paraguay	15
Other Latin American countries	33
Regional total	2,925
Middle East	
Israël	173
Iran, Islamic Republic of	100
Egypt	78
Turkey	65
Lebanon	42
Saudi Arabia	42
Iraq	30
Syrian Arab Republic	16
Kuwait	16
Other Middle East countries	34
Regional total	596
Caribbean	
Porto Rico	114
Jamaica	18
Other Caribbean countries	51
Regional total	183
Africa	
Zambia	51
Nigeria	50
Algeria	50
Kenya	27
Tunisia	24
Socialist People's Libyan Arab Jamahiriya	15
Other African countries	58
Regional total	275
South-East Asia and Oceania	
Philippines	185
India	182
Hong Kong	175
Singapore	142
Taïwan	116
Thailand	98
Malaysia	92
Republic of Korea	92
Pakistan	86
Indonesia	82
People's Republic of China	66
Other countries of South-East Asia and Oceania	65
Regional total	1,381
Eastern Europe	
U.S.S.R.	9,189
German Democratic Republic	759
Yugoslavia	610
Czechoslovakia	466
Poland	340
Hungary	274
Romania	147
Bulgaria	99
Regional total	11,884
World total	136,680

Source: International Data Corporation in *Futuribles*, 1984.

taxation). By automating administrative procedures, State agencies and major firms have been able to acquire the practices and standards of the world economy but, unlike the experience of the industrialized countries, this type of computer use has not created any new impetus in other sectors. The hiatus between the collection and analysis of economic data explains why it is that computers have not by themselves led to significant applications in forecasting, planning and economic management. Moreover the establishment of data banks for the use of developing countries has been largely carried out either by foreign private companies or by international organizations (FAO, WHO, etc.).

Telecommunications undoubtedly provides the most efficient way of decentralizing computing capacity. However, the installation costs of national networks of computer-based telecommunications and the poor quality of existing telephone systems in many countries are responsible for the difficulties that have been encountered in efforts to disperse major computing capacity. Only a few institutions — public or private — have been able to set up their own networks (airlines, banks, etc.).

Starting in the 1970s some developing countries have laid down policies on information technology aimed at reducing their dependence on foreign expertise and at developing their own production capacity. These policies have dealt simultaneously with the technical, industrial, academic and regulatory aspects, the control of imports, and support for local enterprise, leading to the creation of a full range of equipment and systems. Among these countries, Brazil and India were the first to institute such policies and to become exporters of computer products. In 1984 the Brazilian computer industry, with a turnover of US$1.5 billion, admittedly accounted for less than 1% of the world total, but its output nevertheless represents 75% of the total for Latin America and more than 59% of the total for the developing world.

The Brazilian effort dates back to 1972, when a committee was formed to coordinate the design and construction of the first Brazilian computer by the two universities of Sao Paolo (hardware) and the Catholic University of Rio de Janeiro (programming). Two years later the government decided to launch a national computer industry, and the firm of Digibras was set up with two-thirds of the stock in Brazilian hands (a third held by the State, a third privately owned), the remaining third by a foreign company, Ferranti. Digibras in turn created the firm of COBRA (Computadoras Brasilerias), which supplied medium and small computers for government and industry. Digibras itself evolved into an agency for evaluation and encouragement of R&D, and has become a holding company involved in the manufacture of components, the back-up for a national network for data transfer, and the regulatory body controlling contracts for the acquisition of equipment by the federal government (de Melo Teles, 1985; Sultz, 1984).

The policy was confirmed and codified by a law defining those markets for mini- and micro-computers that were to be 'reserved' for Brazilian firms, i.e. those whose capital is held entirely by Brazilian interests, thus excluding henceforth not only subsidiaries of multinational companies but also firms with a mixture of Brazilian and foreign capital, even if only the minority share. Brazil was able to invoke the Japanese and also the American precedents (the Buy American Act) to justify this protectionist policy, since articles 18, 20, and 21 of GATT authorize such measures in two instances — for reasons of national defence, and to protect infant industries.

The result is that the national element in the Brazilian computer industry has been considerably strengthened relative to the predominance of foreign suppliers in the field of big machines, especially American (IBM above all) but also Japanese and French. Between 1980 and 1984 the share of locally produced micro- and mini-computers in the national market has grown from 17% to 95%. The price differential, for the same quality of machine, between the Brazilian products and those of the advanced countries has been diminishing steadily, though it nevertheless remains large (between 13 and 20% for an American personal computer), resulting in a substantial black market in smuggled imported products.

Nonetheless local firms are responsible for 46% of domestic production, which puts Brazil in third place among countries that supply their own needs in this area, after the United States (90%) and Japan (54%). The net result is that the local industry employs on average twice as many people as the subsidiaries of multinationals, and seventeen times as many in R&D activities. The law on information technology, which enunciates the necessity of protecting certain markets in this sector for strategic reasons that are as much military as economic, thus underpins an internal growth factor that stimulates the modernization of industry, a development of skills that strengthens the technical infrastructure and the expansion of the universities, and finally a more 'equitable' share in international trade by favouring the export of manufactured goods rather than of primary products (Erber, 1985)

The majority of developing countries could not hope to adopt a similar strategy aimed at creating the conditions for a relatively independent industrial development. Lacking the size (hence a domestic market of adequate proportions), the financial means, the university organization, the specialist human resources, they are condemned either to import directly the 'black box' of up-to-date technologies, conceived and produced in the industrialized countries, or else to assemble or sell products made from different components manufactured elsewhere. For Brazil the mastery of production appears as a guarantee not just of political independence, but of access to the technologies of the future: hence the description by its president of SERPRO — the agency responsible for data processing at the federal level, simultaneously tax-collecting body and foremost customer of the national computer industry, as well as one of the major centres for training in computer skills — as 'the laboratory of the future' (de Melo Teles, 1985).

The extraordinary expansion of microprocessors on the one hand and the development of space technologies on the other has, however, somewhat modified the situation for many developing countries, making it more viable for them to pursue policies of producing and above all exploiting information and computer telecommunications technologies that are both less costly and better adapted to local resources and needs. In the past, developing countries have tended to underinvest in telecommunications infrastructures (as little as 0.3% GDP), on the assumption that they contributed little to the poorest countries; these views are now changing and investment is increasing (Hudson, 1983). As to computers, some countries have specialized in the production of parts (components, circuits, etc.) for assembly for the multinationals; several countries have established their position in the international market for micro-computers (Brazil, India, Republic of Korea, Malaysia, Taiwan, Indonesia, Singapore) and have set in motion policies on telecommunications based on the launching of their own satellites (India, Brazil, Indonesia, Mexico).

Even if over the last decade more countries have managed to gain entrance to the ranks of the producers as a result of the cost reductions brought about by technical progress, that door remains narrow and the number of the 'elect' proportionately very limited. On the other hand, the number of developing countries that are now better prepared to participate in the technologies and technical systems which are more manageable, more accessible, less expensive and more versatile, is today very much larger than it was thanks to benefits now accruing from educational and training policies in the past.

Impacts on employment

The impacts of the information technologies are diverse and widespread throughout the economic, cultural and political system. It would be unwise to discuss the effects in global terms because of the wide range of situations in different countries, though certain general trends may be discerned. Furthermore, the effects are clearly interrelated: there is obviously no clear division between the level of employment, the nature of working conditions and socio-cultural effects — the problem of employment is inseparable from the nature and organization of the work concerned, while the problems of education, culture and leisure in the shadow of the new technologies can no longer be taken out of their macroeconomic context (in particular the burden of greater or lesser unemployment) or dissociated from the nature of working conditions attached to different sectors of activity. The overwhelming trend in the most industrialized countries is the expansion of the service sector in which information activities will play an increasing part. Even in manufacturing, those involved in processing information will be of greater importance than process workers in the traditional sense.

The information technologies, by their very nature, have had much greater impacts initially on the industrialized countries with their large manufacturing and services sectors than in the developing countries, for whom the ultimate effects are still highly uncertain. The following discussion therefore focuses on the experiences of the industrialized countries. The implication is that for these countries, employment and work will have more and more to do with the use of time rather than of matter.

AGGREGATE EFFECTS

Information technologies have potential applications in all spheres of activity, not just manufacturing, computation and office work. The tendency in all areas is to replace human labour (whether involving physical strength or, increasingly, intelligence) with wholly automated processes which often require supervision rather than continuous action, and which usually employ far fewer people in the production of the same or better results. This does not, however, inevitably lead to a fall in aggregate demand for labour, because of a complex interplay of factors:

— new jobs are created in design, construction, supervision and maintenance of new machinery and new processes

— new products and services are created as a result of new technologies, with concomitant increase in employment

— productivity rises in automated processes, leading to higher value added per hour of work and hence the possibility of greater remuneration, which in turn generates an increase in aggregate demand for existing as well as new products and services

— productivity increases may also be reflected in lower prices, which may further stimulate demand

— jobs are lost in automated processes, and in those products and services rendered obsolete by new technologies and products

The question is whether the positive effects will cancel out the negative ones. In theory, information technologies are no different from all the other technologies involving the mechanization of labour, a process of 'creative destruction' that has been going on since the start of the Industrial Revolution with — hitherto — a positive outcome for aggregate labour demand. Inventions such as the automobile, the radio, television, the airplane, home appliances, the telephone and telecommunications, etc. now account for major proportions of the OECD economies, not only through the production of the products themselves, but also through the many linked user and supplier industries that they have generated. These are both service and manufacturing industries. Growth in productivity has been accompanied by an increase in the economically active population in all the industrialized countries: since 1880 employment in the OECD countries has tripled, despite the fact that mechanization has eliminated two thirds of the jobs that existed prior to that date (Geldens, 1984). (It is true that life hours worked also dropped, but in Britain, for example, this has been at a rate a little less than one third as fast as the fall in the number of hours work required for a given output (Williams, 1983a, 1983b).) Total employment in fact has increased most rapidly in those countries which were the first to adopt labour-saving technologies, though the growth has tended recently to be in the service sector rather than manufacturing. Overall, in the last thirty years, among manufacturing industries, high rates of productivity growth, high R&D spending relative to sales, and rapid appearance of new products have had a high correlation with more stable prices and rising employment.

From the outset, mechanization has generated fears about massive unemployment whenever there has been an economic crisis — most recently following the oil shock of the early 1970s which happened to occur at the same time as the most rapid technical change in microelectronics. Although unemployment certainly increased dramatically (10 million unemployed in 1970 in the OECD countries, as against 30 million in 1985), it is far from clear that the expansion of microelectronics was the direct or the sole cause of the rise. The rapid pace of technical change undoubtedly exacerbated the overall rise in unemployment in the medium term, and it certainly contributed to the altered pattern of labour demand (see below). One problem is that it is always much easier to predict the jobs that will disappear as a result of automation than to predict or identify the new jobs that may be created.

Robotics may prove to be a special case, in that it is argued that robots do not simply increase the productivity of human labour, they effectively replace the worker (Shaiken, 1985). Computers have opened up the possibility of the fusion of design, manufacturing and marketing into a single stream of information that will eventually permit us to automate just about anything that we do not wish to do ourselves. Output can be increased and new products created and manufactured without any concomitant expansion in labour demand. As long as the capital costs of robots remain high relative to the cost of human labour, there will be some impediment to the wholesale replacement of human workers, but the trend is already apparent.

According to the forecasts of Arthur D. Little Inc., the market for robots will be worth 2 billion dollars by 1992, with 50,000 new robots coming into operation each year — more than there were in all the factories in the world in 1984. Expansion on this scale could have extremely negative effects on employment, if robotization does not create more jobs than it replaces, or if the service sector fails to create proportionately more jobs, or else if the size of the labour force does not diminish. The pessimism of Leontiev (1985) arises from doubts as to the likelihood of these three conditions being satisfied, and he goes so far as to reckon that in the United States of America by 1992 three-quarters of a million managers and five million office workers could be unemployed. His scenario envisages that automation of office tasks could have an even more negative impact on employment in the medium term than robotization. If this forecast is fulfilled (it is very controversial), even in the short term these developments could have devastating consequences for many developing countries — by not merely failing to increase employment, but by actually destroying it.

STRUCTURAL EFFECTS

The impact of technical change inevitably varies between regions, sectors, firms and skills: jobs are created in new productive activities in innovative firms and industries and are lost in areas and skills that cannot adjust to new processes and products. In the case of the information technologies, there has been a boom in employment in the manufacture of a vast range of new microelectronic products and in the services linked to them, while at the same time large numbers of workers have been made redundant as a result of the introduction of machines that replace their strength or skill and or of new products that render their own output obsolete. The consequences have often not been predictable. The modern information technologies are much less tied to specific production locations by their physical inputs than most conventional manufacturing, and they have tended to flourish where there is a suitable labour supply (a pool of semi-skilled assembly workers — often female — for the hardware, and of highly educated personnel for the software). In some instances this flexibility has benefited the newly industrializing countries (Taiwan, Singapore, Republic of Korea, etc.) and previously run-down areas of developed countries (such as the so-called Silicon Glen in Scotland). In general the result has been to favour the employment of women over that of men, and of white collar over blue collar occupations.

Training (or retraining) is obviously the key to adjustment to changing conditions, though in practice difficult to achieve sufficiently rapidly or fully to avoid a real risk of structural unemployment: the chances of finding a job are much reduced for unqualified young people, for older and less mobile members of the work force tied to production and sales of traditional goods, for women looking for secretarial jobs and for workers without higher education or qualifications. In these categories, unemployment can be long lasting, even though the supply of jobs as programmers, engineers, specialist maintenance workers, etc. continuously increases in all branches of electronics.

But even training the work force is only a partial answer if, as many fear, the new manufacturing activities and the service sector cannot generate sufficient jobs to compensate for the loss of jobs elsewhere in the economy. Given the generally modest rate of growth in the industrialized countries since the 1970s, the impotence of the conventional remedies in dealing with this situation, and the increase in service sector productivity resulting from information technologies which may mean that employment growth in the sector is very limited, the ultimate solutions may lie in work sharing, reduction in hours worked and radical transformations not just of the structure of employment, but of the way that work is organized, the institutions and values of economic life (Freeman, 1982; Leontiev, 1985). It is possible to conceive of technical progress modifying the very nature of what constitutes 'work' and 'leisure' as it creates activities and occupations increasingly unlike traditional productive tasks.

INTERNATIONAL DIVISION OF LABOUR

Technical advances in information technologies in the industrialized countries have inevitably had repercussions on the rest of the world and will continue to do so. There are two aspects to this impact: on the evolution of the developing countries' economies (i.e. on the process of development itself) and, intimately linked with that, on the role that newly industrializing countries can play in the international economy, in competition or cooperation with the industrialized nations.

The diversity of the Third World of course means that there is a very broad range of experiences and of potential. As to the evolution of the economy, the stage of development already reached is obviously relevant, though it is not clear whether the newly industrializing countries will be obliged to pass through the same stages of development, and in the same way, as the industrialized countries, or whether their circumstances are so different that their experience must also be very different, especially given the possibilities offered by the new technologies of bypassing some of the steps. The new information technologies undoubtedly provide a way of speeding up the pace of skill-building and of structural adjustment of production to conditions in the international market. In view of the importance of the information technologies for the service sector, much depends on whether the country in question can expand its service sector or whether the other sectors, particularly agriculture, remain dominant; the realistic assessment must be that large-scale growth in the services sector is likely to take a very long time to happen.

It is difficult to foresee the ultimate impact of current developments on the international division of labour. Until recently it was thought that the developing countries, particularly the newly industrializing countries, would gradually take over the mass-production of standardized manufactured products, leaving advanced technology, customized, high value-added products, as well as design and engineering services, to the developed countries; (Sabel, 1982; Telesis, 1982: especially ch. 1, pp. 44-62). The basic argument is that as dominant product designs and production technologies become established they are readily transferable to developing countries, where drastically lower wages and less social protection of workers enable production to be carried out at much lower cost — more than enough to offset transportation costs and lower labour productivity. Such a trend was readily observable in the 1970s, when many American and European firms exported labour-intensive assembly operations in consumer electronics, semiconductors, and ultimately industrial electronics and computers. It was a period also when automobile companies were subcontracting to foreign suppliers of components. Indeed the Ford Motor Company had announced a strategy for a 'world car' with components and assembly in many different countries according to the location of comparative advantage (*Financial Times*, 16 November 1984). In the semiconductor industry, where assembly turns out to be one of the most labour-intensive of all manufacturing operations, off-shore assembly in foreign affiliates grew at a remarkable pace. It was most striking along the border between Mexico and the United States, where low transportation costs made it an especially attractive option (Grunwald and Flamm, 1985).

Even with the introduction of new robot production technologies, it has been pointed out that these technologies, when fully exploited, will tend to reduce the importance of scale economies, and therefore permit much smaller production volumes to be located in developing countries without cost disadvantage (Sabel, 1982; Hirschhorn, 1984; Ayres and Miller, 1983). Moreover, longer working hours and the willingness of workers to do shift work make possible higher utilization of production capital, and thus save capital as well as labour costs. Modern transportation and the combination of computers with telecommunications make possible the close coordination of production at widely dispersed sites, and the easy transfer of design to production via computerized techniques.

The preceding argument is now being increasingly called into question, however, supported by fragmentary evidence for the repatriation of production to developed countries, using highly sophisticated programmable automation and flexible manufacturing systems (*Newsweek*, 12 March 1984; *Electronic Business*, Feb. 1984; *Fortune*, 11 July 1984). As the more labour-intensive phases of production, such as assembly, become automated, the argument is that the labour content of most segments of manufacturing will become so small that labour costs will cease to be an important factor in plant location decisions and the developing countries will cease to enjoy, if one can use that term, the cost advantages of having a cheap labour supply. Other factors, such as the costs of inventories and buffer stocks, the advantages of being close to final markets, and of close collocation of suppliers and assemblers so as to take advantage of close interaction between design and production and 'just-in-time' inventory management systems, increased skill requirements for the residual work force that is still required to avoid breakdowns and keep production running smoothly — all of these will tend to outweigh labour costs in location decisions (Junne, 1984a; Jacobsson and Sigurdson, 1983; Edquist and Jacobsson, 1985). The replacement of economies of scale in production by 'economies of scope' — the use of modern flexible manufacturing systems to produce a wide variety of components or even complete product lines with the same capital, flexibly programmed — will give the advantage to close collocation of all the elements of design, production, and marketing where all these elements can be integrated. Although branch plants already located in developing countries are not likely to be closed soon, future plant expansion and the production of the newer generations of products are likely to take place in the developed countries, leading to the stagnation of the export-led growth which was so successful for many newly industrializing countries in the 1970s.

In addition, the widespread adoption of biotechnology techniques in many aspects of chemicals and materials production will tend to reduce the demand for primary products from developing countries, thus further weakening their position in world markets (see Chapter 3).

At the present time, it is difficult to sort out these arguments. For example, many recent studies of the introduction of programmable automation and flexible manufacturing systems have indicated that the theoretical flexibility that these technologies offer in increasing the variety of products that can be produced with the same capital equipment is very difficult to realize in practice. At a minimum a radical change in management philosophy and the organization of work is required before the economic benefits can be realized (Fadem, 1984; Kelley, 1985). A recent study of almost all the flexible manufacturing installations in the United States and Japan indicates that the average number of different parts turned out per American installation was 8, while in the average Japanese installation it was 30 (Jaikumar, 1984). This experience may offer either a warning or an opportunity to the newly industrializing countries. The apparent achievement of much greater flexibility in Japan suggests that, in part because of their more recent industrialization, they have been able to adapt organizationally much more readily to the new manufacturing technologies because they are less bound by inheritance from the older traditions of 'scientific management'. The developing countries may similarly have the opportunity to make the necessary organizational and work force adaptations which would enable them to 'leapfrog' into modern manufacturing with smaller batches and greater product variety. It is really too early to tell. Furthermore, the difficulties experienced in the adoption and implementation of programmable automation may turn out to be only 'growing pains' that will be overcome in the longer term as both the design of the technologies and of the manufacturing organizations evolve. If this learning occurs more rapidly in the industrialized countries, the closeness of plants to both customers and designers may provide an overwhelming comparative advantage which will be reinforced by the new technologies.

TRANSFORMATIONS IN THE NATURE OF WORK

The effect of the information technologies on total employment can vary considerably between the developed and developing countries. On the other hand, there is no reason why the qualitative effects on skill requirements and working conditions should not be similar: the service sector is increasingly going to involve tasks linked to use of a computer terminal, and the industrial sector will be marked by some degree of reduction in the role of physical effort, accompanied by an expansion in activities of supervision and control.

The new technologies affect the nature of both the tasks themselves and the working environment, as well as the overall level, distribution and conditions of employment. Many of these effects are associated with the changing organizational structures resulting from the introduction of the new technologies. In fact, information technologies play a multifaceted role in the functioning of an organization, being part of the communications network, of the management process (decision-making, coordination, control), as well as of production itself (via automated systems, word processors, etc.).

Nature of skills

As the use of new computer technology in manufacturing has spread to more industrial work places, the issue of the impact of technology on the nature of worker skills in the industrialized countries has once again become the focus of considerable research, academic debate and popular speculation. Most discussion tends to fall into two broad categories. The first (which might be termed the 'upgrading' hypothesis) argues that, while technology obviously renders many traditional worker skills obsolete, it requires new and substantially more complex skills on the part of the manufacturing work force. From a long-term 'macro' point of view, this hypothesis appears historically plausible because it is obvious that the composition of the work force in all industrialized countries has been shifting towards a higher percentage of 'professional and technical' workers, and of skilled workers generally. According to another point of view (the 'deskilling' hypothesis), new technology tends to bring about a progressive loss of skill on the part of manufacturing workers as skills are redistributed not only from worker to machine but also from production worker to managerial and technical personnel (Bright, 1958; Braverman, 1974). There is some truth in both lines of argument, but they are too simplistic and do not take account of the complex interaction of a wide variety of other factors involved, in particular management attitudes and practices.

Recent case studies have shown that traditional skills continue to be very important, even in highly automated plants. Human intervention is still required to sort out problems when something goes wrong. Programming and debugging of software has proved to be much more complex than most engineers and managers expected (Kelley, 1984), and there can be problems for quality control because of the lack of natural pauses in the work flow (Blumberg and Gerwin, 1984). Skill in handling the problems of computerized manufacturing technologies 'results from experience on the machine, and increased emphasis needs to be placed on the maintenance and increase of craft or shop-worker skills' (Nicholas et al., 1983). There have been similar experiences in process industries such as chemical plants, nuclear power plants or oil platforms (Fischer, 1981; Fadem, 1984). If managers assume that skills are no longer necessary when new technology is introduced, they may be unable to realize the goals of increased productivity that the technology is supposed to make possible.

There are potential problems, too, for training if there are no longer opportunities for people to acquire the skills on the job that are crucial to efficient production. Inexperienced workers have little chance to observe more skilled workers in action, and even less chance of practising general machining skills first-hand in a situation that poses no hazard to the overall process (Kelley, 1984).

Computerized manufacturing systems also place new kinds of demands on production workers, calling for new kinds of skills in addition to those of the traditional machinist. Whereas before workers were responsible for only one aspect of the production process, they are now — at least partially and indirectly — responsible for the entire production system (Adler, 1983, 1984). The greater complexity and integration of the technical systems, and the way that work now has to be organized, requires cooperation and interdependence among workers at all levels (OTA, 1984: 193). Workers have to develop more abstract skills, to monitor and maintain systems rather than operate them (Adler, 1983). Engineers argue, however, that these new demands are merely transient, the result of 'growing pains' in the new systems; and that they will diminish as design and operation improves.

The more specific problems of skill displacement, which always accompany technical innovation, have been mentioned above. The pace of technical change is now so rapid that it is already making obsolete skills that were considered highly desirable less than a decade ago. Until recently, for instance, the diagnosis and repair of computer breakdowns required a very thorough knowledge of the working of the machine, and hence highly specialized technicians. Nowadays, the computers have built-in circuits that are designed to monitor their functioning as well as programs that can make relatively precise diagnoses of problems. Technicians have moved from spending all their time looking after a single machine to working over a much wider geographical area, controlled from a central headquarters; with the latest developments in this area, which make possible remote-diagnosis of problems, the number of technicians concerned with diagnosis diminishes at the same rate as the number of repairers, since technical progress reduces simultaneously both the number of faults and the duration of the repairs.

Management attitudes and techniques

The new technologies and worker skills do not accord easily with the principles of 'scientific management'. The traditional view, going back to Frederick Winslow Taylor almost a century ago, is that efficiency is best served by an ever more refined and hierarchical division of labour, with the decomposition of tasks into ever more fragmented elements; factory work is to be designed according to well-established 'scientific' principles and sustained by close supervision supported by quantitative measures of worker performance; management must have exclusive control over design and introduction of new technologies, investment and location decisions; workers' interests are best protected by elaborate work rules, well-codified grievance procedures and detailed written collective bargaining agreements.

The validity of this approach is still widely accepted among manufacturing managers in industrialized countries, although it has recently been under attack from a number of standpoints. Abernathy and Hayes (1980) have criticized the results as having significant efficiency, as well as social, costs; workers in industrialized countries, now more educated and accustomed to political democracy outside the work place, no longer respond to the authoritarian hierarchies and tight disciplines of the so-called 'technocentric' approach. Others argue that scientific management is simply not able to cope with the needs of flexible production systems or of the fast-changing, unpredictable economic and technical environment in which most firms must operate (Sabel, 1982; Piore and Sabel, 1984).

The difficulty of changing entrenched managerial attitudes (Skinner, 1983) is compounded by the fact that much of the thinking of organized labour, too, has been conditioned by the Tayloristic model of production. Precise job descriptions with related pay agreements and codified sets of work rules have traditionally provided the worker with protection against arbitrary and 'subjective' management decisions, but at the same time, they have retarded adjustment to changing markets and technology. Taylorism thus provides different but complementary benefits to both labour and management which make it difficult to alter the system without clear practical demonstrations of the benefits of alternatives. Nevertheless attitudes on both sides are beginning to change, and technology is becoming a crucial bargaining issue. Recent examples of concern in this area include the 'Technology Bill of Rights' drawn up by the International Association of Machinists and Aerospace Workers (1981); the technology clauses in agreements between AT&T and the Communications Workers of America (1980); and legislation regulating the introduction of new technology, with active worker participation, in Norway and Sweden (Schneider, 1984). Major changes in approach to this issue are hampered by management reluctance to cede control (since there are clearly much wider repercussions) and by lack of expertise on the side of organized labour.

New technologies have been welcomed by management as a means of increasing control over the work force and reducing vulnerability to strikes (since operations can be continued using supervisory personnel), though they often overestimate the ease and rapidity of the introduction of new methods. Recent historical research has suggested how goals of maximizing managerial control with minimal human input at the working level have been an important driving force in the development and adoption of computerized manufacturing (Noble, 1984). In fact, managers will have to understand that, rather than a means to expand management control in order to carry out production with less skilled workers, technology must be understood as just the opposite: as a means of extending the capabilities of all workers so as to increase the flexibility of the production system and its capacity for responding to rapid, and often unpredictable, changes in technology and in the market environment.

Organization and quality of working life

It is a common assumption that new computer-based technologies provide a unique opportunity to restructure work so as to enhance the work environment and improve the quality of working life, eliminating arduous and repetitive tasks. One of the clear advantages of automation is that it can improve health and safety by eliminating dangerous jobs or jobs in unhealthful or unpleasant environments, and by increasing the distance between workers and the production process; for the firm there are consequent savings in environmental control and safety device costs to protect human workers (in Japan this has been the primary incentive to instal robots: Toda, 1984).

However, the effects have not proved entirely beneficial, and in particular, some new psychological hazards have been introduced (OTA, 1984). These are in general much harder to cope with than purely physical problems because they vary more widely among individuals, depending on educational levels, social background, and the nature of the working relationship with the machine. Psychological and social stresses are usually associated with two factors: (a) responsibility for complex, highly integrated, expensive, interdependent systems, and (b) computerized monitoring of individual worker output and pacing by machine. Health and stress problems are significantly influenced by the nature and organization of the work. It appears that where technology is used to augment the intellectual or physical powers of an individual, the quality of working life is improved; but where it is used to mechanize low-skill work, making tasks more monotonous and repetitive, psychological stress is increased (see, for example, the case study of differences in stress of professionals and secretaries using VDTs: Smith, 1980). There is often a conflict between the infrequent necessity for very sophisticated intervention in complex systems combined with long periods of routine operation with little for the operator to do: high attention demand and responsibility in combination with little action can be very stressful.

Often the social system in the work place is expected to adjust to the technical design, and not the reverse; nevertheless adverse effects can be modified according to the strategies used to design, introduce and implement new technologies — these choices can be influenced by managers and workers as individuals, and by workers through their unions, both formally and informally. The success or failure of new technologies ultimately hinges upon the way these problems are handled, more than upon purely technical factors. Technical factors may underlie the pace, discomfort, danger to health, psychological pressure, risk of accidents, etc., involved, but these factors are not in themselves entirely responsible for the way that technologies are used in the daily operations of the workshop, factory or office. The machine of itself is not oppressive; what matters is the state of relations, controversies, respect for existing legislation and negotiations within the firm, as well as the degree of cooperation between technicians and operators.

A desirable work environment, based on 'human' criteria, should include some of the following features (Maccoby, 1984; Walton, 1985):

— respect for individual differences, and the realization that the attitudes, expectations, needs and values of individuals in the work place differ and that it is necessary to match work tasks to the characteristics of individuals, who cannot simply be treated as though they were interchangeable parts in a machine;

— good wages and benefits, usually defined as above-average for the economy as a whole, but in any case involving the equitable sharing of the economic rewards of productivity gains;

— employment security (in contrast with security in the performance of a particular set of tasks) in exchange for high mobility of workers among tasks and job functions within a firm, and a willingness to learn new skills and accept new challenges;

— a work place free of any safety and health hazards appreciably greater than those experienced by the general population or by workers generally;

— interest and challenge in work, with emphasis on tangible work products for which individual workers or teams can accept responsibility, and in which they can take pride;

— personal or team control over the pace of work, with the opportunity to make independent decisions within broad guidelines about how work is to be done, leaving workers with discretion as to when or if it is necessary to obtain authorization from higher management;

— recognition within the organization for superior group or individual performance, a sense of appreciation by management and peers which transcends merely economic rewards;

— opportunities for personal growth, learning, and career advancement, with the range of skills continually expanding, and payment based on acquired competence rather than particular transitory job assignments;

— good personal relations with peers and superiors, based on mutual respect, and on absence of non-functional status differences and authority symbols, minimum number of layers of management;

— a significant say by workers or their representatives in the design, pace, and mode of introduction and implementation of new manufacturing technology, particularly as it affects the working environment and the allocation of tasks and skills among workers.

Ideally, the design of manufacturing systems should meet both human and technical economic criteria, i.e. a humanly satisfying work environment combined with economic efficiency. In the traditional Tayloristic philosophy these are seen as inherently conflicting requirements; however, an increasing number of observers are now suggesting that the human and economic criteria are compatible and, indeed, mutually reinforcing. Nevertheless, it is not clear that changes in this direction will take place as a result of competitive market forces: the question turns on whether social costs are automatically 'internalized' in the prices of goods and services or whether sociopolitical intervention is required to bring this about.

How does the experience with technology and the quality of working life in the industrialized countries affect the prospects and policies of the developing countries? To some extent, of course, present conditions in the developing countries are more nearly like those that existed in the early days of industrialization in the West. The industrial labour force is recruited largely from rural migrants with low expectations and levels of education, for example. The developing countries have a need to mass-produce standardized inexpensive products with imported manufacturing technology; the pressures on their industrial work forces will be much the same as in the developed countries. Once again, the crucial factor will be the way that new technologies are handled. The flexibility that will increasingly be inherent in imported production technology that is only a little behind the current state-of-the-art offers the opportunity to avoid many of the less attractive aspects of the technocentric approach, and to experiment with newer forms of work organization appropriate to the social systems and cultural traditions of the countries in which they are being introduced. There are scattered experiences in the developing countries which demonstrate that the ideas about improving the quality of working life being tried out in the industrialized countries can be implemented in developing countries without sacrificing productivity or efficiency. Not all the desirable aspects of new approaches to management require interaction with a highly educated work force.

The spread of information technologies will certainly hasten the arrival of the high-productivity society, with jobs, working conditions and attitudes to work that are very different from those associated with traditional industrial life in the developed countries. In this area, yet again, the circumstances of most of the developing countries are very different from those in the industrialized countries. The activities that handle information as the basic input, for manufacturing or services, are the province of an expanding white-collar work force. For the developing countries, the growth of these activities will be limited to a tiny minority of the active population for the time being. There is surely a risk that the better working conditions and pay that this minority enjoys will accentuate the 'dual' nature of these societies, where traditional activities will predominate for a long time to come.

Computerization prolongs and accentuates the technological duality discussed in the introduction: at one and the same time the public scribe in the street or market and the word-processor in the administration, the army, banking and insurance, industry or the university, each plays an essential role for all that they relate to worlds that are widely separated in time. In the industrialized countries, many commentators have been exercised by the unequal distribution of opportunities and advantages that the new technologies provide, fearing that the result will be a growth in that part of the labour force condemned to arduous working conditions at low wages, without any hope of social improvement. This unfair distribution is all the more apparent in the developing countries. The risk of a proportionate increase, rather than reduction, in the part of the population unfitted to work with computers is all the greater: the gap between a technician class that benefits from the better working conditions and all the advantages of the electronic revolution and the masses — handicapped in the race to progress, condemned at best to the laborious tasks of a past era if not to unemployment — could continue to widen.

Cultural and political stakes

Knowledge is power: Bacon's dictum is even more valid for the technologies related to information and communications than to those of the past based on energy and materials, because these new technologies mechanize functions hitherto the province of the human mind, by accelerating, multiplying, broadening its scope to a prodigious extent. There are indeed direct consequences for power in the political and military sense — in a modern war, however 'conventional', fought with sophisticated weaponry, the strength of C^3 (command, controls and communications systems) would count for much more than the physical might of the arms themselves. But the social and political implications are more fundamental than the mastery of the individual information technologies would suggest, and involve a radical change in society, perhaps of our whole civilization, that affects attitudes of mind, behaviour and values. This new knowledge is all the more powerful because it concerns the *collective* intelligence, language, memory, culture, organization, not just that of the individual or small group. A two-way process of adjustment is necessary: society must adapt to the computer, but at the same time the computer must be adapted to society.

Technical breakthroughs have made information infinitely more accessible — computers and telecommunications in theory allow instant access, anywhere on Earth, to the global stock of organized knowledge, constantly updated. The parallel with the introduction of printing is obvious, and — just as in the case of printing — the new technologies are causing a truly cultural revolution. The transition from the written to the printed word was not simply responsible for popularizing reading, it provided science with new means of standardization, of storage, of propagation, and hence of discussion and research.

'One cannot treat printing as just one among many elements in a complex causal nexus, for the communications shift transformed the nature of the causal nexus itself. It is of special historical significance because it produced fundamental alterations in prevailing patterns of continuity and change' (Eisenstein, 1979: 703; see also pp. 698, 700-1).

Just such alterations, but on a much more spectacular scale, are being generated before our very eyes by the new information technologies, which immeasurably enhance all the advantages brought about by printing and in addition open up completely new capacities to master Nature through the processing and analysis of information.

The virtually limitless range of the technically possible is, however, subject to major impediments of cost and control which operate particularly heavily against the developing countries and threaten to accentuate the growing socio-cultural rift between the developed and the developing nations. Computers and telecommunications provide the most efficient method ever invented of circulating information and making it available to all, but they also allow information to be priced, and in particular make it possible to charge for information that was previously free. Memories relating to traditional craft methods, for example, that have hitherto been freely available, have been replaced by computerized systems that are more costly and difficult of access, since specialized equipment is needed for retrieval. The costs may be prohibitive for developing countries: barred from access, they will be held back even further.

Cost is not the only problem. Data-banks are not simply more expensive than books (even though they make the best libraries mobile and accessible anywhere in the world), their contents are far from being neutral. The supremacy of American firms in the area of bibliographical data-bases and their dominant position with regard to data-banks carry a risk of intellectual alienation that the developed countries are the first to point out: might not the reorganization of knowledge around this new method of classification, formulation and recreation of the collective memory lead to a loss of identity? The vision of the world that the *New York Times* data-bank presents is not the one that nations other than the United States may have of themselves; but if these countries are obliged to seek their sources from that data-bank, what image are they going to have of their own history? (Crémieux-Brilhac, 1980)

A fortiori, such a dependence seems inevitable for the developing countries who cannot set up their own data-banks and have to rely on foreign firms and memories. More than 95% of scientific and technical information is now held solely by the industrialized countries, while many developing countries have to purchase abroad information pertaining to their own national circumstances. Furthermore, the pool of technical information (agricultural, medical, socio-cultural, socio-economic, etc.) on which the developing countries must now draw is based overwhelmingly on the experiments and the techniques developed in the context of the most advanced industrialized nations.

Information technologies are the product of a culture that may acquire global domination at the price of cultural impoverishment for many countries, if not for all of humanity. So far, too, they have been linked to a single (Western) script: behind the supremacy of the programs and the language structures that are associated with these models, there lies a mode of thought, a style of economic and social organization, and a cultural context the roots of which are alien to the socio-cultural situation of most developing countries.

The debate about the threat to cultural identity posed by the new technologies is not recent: it started and grew with radio, the cinema and most especially with television. Microelectronics — less costly, more flexible and versatile — could lead to a diminution in cultural dependence, provided that the countries concerned are better placed to devise and produce the programs that they need. However, the link between computers and telecommunications may well set the debate off on a new round since satellites now permit the instantaneous direct transmission of visual images, a development which promises to have an even greater impact, because immediate. The growing interpenetration of telecommunications, computers and audio-visual technologies carries a cultural cost, and those countries that are unable to produce their own independent broadcasts, films and cassettes risk daily bombardment with the products of foreign media networks, and hence are both dispossessed of their own culture and subjected to the influence of news bulletins from foreign sources.

Information cannot be separated from its organization, i.e. from the way in which it is stored and distributed. The control of transborder data-flows and data-banks would appear to be a requirement of sovereignty, although national controls could conflict with the principle of free circulation of information — freedom of expression and freedom of information exchange. In this regard the cultural stakes cannot be separated from the political stakes. The public authorities (the legislative and judiciary as well as the executive branches) must define new ways forward between concentration and dispersion, between excessive control and *laissez-faire*, between cultural autonomy and domination, between concern for the rights of the individual or of privacy and the defence of national interests.

Many of these problems are so recent that it is still too early to say whether they are real or imaginary, and how serious and urgent they are. The analysis of the international repercussions of the development of computer-linked telecommunications cannot be isolated from that of other developments, such as the controversies about the new international division of labour or 'the new world order'. It is hardly surprising if behind the cultural debate there lurk tensions associated with economic struggles and power conflicts: where the OECD countries talk about freedom of information and free trade, developing countries talk in terms of the drive towards domination and neo-colonialism based upon technological superiority; where the former seek to open up the market so as to allow the transfer of technologies, the latter insist on some degree of protectionism in order to permit the emergence of production of their own technologies.

Transborder data-flows are growing at the rate of 15 to 20% per annum, but that figure indicates only the volume of transactions and not their content and nature. This last is obviously the most critical. No customs control can be imposed without interfering with the privacy that protects all correspondence: magnetic tapes can be taxed, but not the information that they contain (and their value may derive not from their actual contents so much as from the expectations of the recipient). Furthermore, the new technologies directly threaten the public monopoly over post and telecommunications: the reduction in the cost of satellites, and especially of the land-based antennae linked to them, is likely to lead to a proliferation of private transmission networks that could saturate whole regions without regard to local sovereignty.

This is but one case among many illustrating how the technological factor impinges upon economic ideas and practices: the mood is towards deregulation in the 1980s as a result of technological pressure as much as of economic, political and ideological changes on the part of policy-makers and the general public. The traditional frontiers between the public and the private spheres (whereby collective interest has led to State intervention for the production and diffusion of technical innovations are challenged in the most developed countries by the very evolution of technology, calling for more decentralized initiatives and less control exerted by State bureaucracy on the management of large sectors such as telecommunications, TV channels, transportation.

These technical battles have immediate political dimensions, as is the case in the share-out of wavelengths. In the course of discussions with the ITU (International Telecommunications Union), the South complained that the North had benefited from the principle of 'first come, first served' so that there is no room left in certain frequency bands, yet to find less crowded wavelengths (because at a higher frequency) requires access to the latest technologies which are precisely those beyond the reach of the developing countries. Similarly for orbiting satellite stations, which provide the best possibility for developing countries to establish their domestic telecommunications networks — the best positions are already taken.

The rapid spread of telecommunications networks, in particular the satellites that can transmit simultaneously sounds, images, texts (in code, multiplex, etc.), opens up an enormous area to transborder data-flows, and at the same time it generates a whole new order of problems for the policy-makers to deal with.

Already on the national level, for instance, the political authorities cannot avoid intervening to define the 'rules of the game' since what is at stake is both the protection of privacy and the guarantee of democratic principles. If access to the data-banks is limited to a small privileged group, or perhaps merely to the immediate powers-that-be, there is obviously an imbalance that could lead to abuses: the totalitarian tropism of computers is not just an Orwellian fantasy. If, on the contrary, access is totally unregulated, there is a danger that confidentiality of some information could be jeopardized, there being new dimensions to the temptation to leak information for reasons of trade, industrial or political espionage but also fraud or blackmail.

The problem of the protection of privacy is in fact what has led the industrialized countries (Sweden first, in 1973) to devise new legislation relating to computers, and to try gradually to harmonize their laws through multilateral agreements. There are two reasons why it is particularly difficult to bring laws into line in this area: the rapidity of the technical change that underpins the circulation of data internationally, and the diversity of juridical and cultural traditions. The suspicion and resistance that any particular project of legislation may arouse is understandable — the basic rules are not necessarily the same in every country or group of countries, and a given specific barrier (for instance, the protection of individual liberties) can easily hide completely different purposes (economic protectionism, for example). If suspicions of this nature exist between the developed countries, it is not surprising that they cause deeper divisions between the representatives of North and South in international arenas. A major achievement in this area is the adoption by OECD in 1980 of General Guidelines, which, though not binding, have been widely endorsed. The same principles have also inspired the Convention signed by the Council of Europe in 1982, which is binding (OECD, 1985).

Unlike the other technologies, the information technologies involve intangibles; the attendant problems are therefore of a completely different kind and may be on an utterly different scale than hitherto, making the computer-based society potentially infinitely more vulnerable to the risk of catastrophe, accidents and criminal acts (computer fraud, illicit access to distant data-bases, etc.).

The evolution of information technologies and the problems that they generate at every level have led all the industrialized countries to establish interministerial committees, and sometimes specialist agencies, to anticipate developments and to prepare responses to the issues that will be raised. There are similar organizations on the international level, for example the Information Computer Communication Policy Committee (ICCP) of OECD. The Torremolinos Conference on the subject of 'Strategies and Policies for Computers' indicated that the developing countries would do well to follow the same course of action, to organize and equip themselves, first of all taking expert advice in order to be in a position to understand the implications — industrial, economic, political, social, juridical, cultural — of the rapid and pervasive spread of information technologies, and to devise appropriate policies, either independently or else in cooperation with other countries, developed or not.

Some issues

The developing countries have little choice about *whether* the information technologies are or are not going to affect their development strategies, the question is rather to know *how* they are going to keep the problems of the sudden technological invasion to a minimum and at the same time make the most of the possibilities that are offered. The way that the industrialized countries have faced these same questions could help in identifying and illuminating the problems at issue, although the situations are so very different that the same conclusions should not necessarily be drawn.

Even the least industrialized settings can now be reached thanks to developments in miniaturization and telecommunications, which have a bearing not only on the implementation of very advanced systems but also on the improvement (cost reductions and greater efficiency) of more conventional systems: independence from mains electricity supply, easier operating conditions with more 'user-friendly' access, and finally the growing link between computers and telecommunications via satellites which will allow systems to bypass the inadequacies of the telephone network and overcome physical difficulties of terrain, all of which suggest that there may be new fields of application that are not exclusively delineated in terms of the needs of the industrialized countries.

The improvement of data-banks and access by means of satellites, for example, makes it easier for small communities that lack substantial infrastructures to make use of the information technologies. The installation of decentralized systems could have beneficial effects not only for the more efficient management of small and medium-sized concerns (rural cooperatives in particular) but also for agriculture, health and education. In fact the assessment of natural resources, the spread of meteorological data and forecasts, the prevention of certain natural disasters (for example by anticipating the migration of swarms of insects) could gain from the flexibility of new systems of data collection and processing. Several developed and developing countries are already cooperating on projects involving the distribution of drugs, health education and delivery systems, epidemiological surveys and assistance in medical diagnosis via access to expert systems (see Chapter 4).

On the other hand, few developing countries are in a position to adopt the new information technologies on a large scale. Even if some of the new techniques bypass deficiencies of infrastructure, others are closely dependent upon the existence of sophisticated facilities and organizational structures. Even if the tendency is towards non-language communication with machines, the vast majority of computers and word processors require literate operators (and literate in Western languages at that). Besides which, in spite of falling costs, the investments involved are necessarily large in relation to resources so that financial considerations alone will limit what is feasible; in many areas, other more basic needs must be given higher priority than new information technologies.

The choices before the developing countries are all the more critical, given that the industrialized countries do not share these impediments and stand to gain exponentially from the new technologies. The developing countries do not face a static situation where their competitors will stand still: the new information technologies give the industrialized countries a huge head start *and* an accelerating development potential.

More than ever, the emergence and growth of a new technical system is raising the issue of cooperation and competition between nations (with all the inherent tensions, traps, illusions, bottlenecks, and conflicts). Telecommunication and information systems networks are international, both in their hardware and in the interconnections among national networks. Furthermore, information is a recognized commodity of international value in trade and commerce. Moreover, it is an unusual commodity in that it does not lose its value on being transferred or used. And, as already frequently emphasized, it is also unusual in that one cannot easily differentiate the medium from the message, the means of transfer from the content of the information being transferred.

For all these reasons, the developing countries have no choice but to consider the policy that they should formulate, the margin of manoeuvre available to them, the strategy that they must adopt in order to control this new tool that has such immense repercussions on all aspects of their economic, social and cultural life. In this last area even more clearly than the others, the response obviously depends substantially on education and training policies. If mastery over the *production* of the new technologies is related to factors and circumstances that are beyond the control of many developing countries, the mastery over *use* to a great extent depends upon the efforts initiated and resolutely pursued over the long term to train, retrain and accommodate an ever larger section of the work force to working with the machines, languages and applications that these countries *truly* need (Zhang, 1985).

The information technologies are an indispensable tool for development, perhaps one of the shortcuts to 'catching up', but not all of the equipment or the applications are of equal value in the process of development: between the gimmicks that are satisfying to national vanity and the power struggles of the multinational companies to protect or extend their markets; between the protectionism that condemns a country to stagnation by shutting out all technical progress and the temptation to follow blindly the model of the consumer society — the policy on computers and information in the developing countries must find an approach that responds to each country's particular situation. To evaluate and select among the alternative options, to decide eventually the kind of technologies and the different possible ways of introducing them that each country requires is undoubtedly the greatest challenge of all, if the Third World does not want to turn its back on or alternatively to submit passively to the information revolution.

References

ABERNATHY, W.J.; HAYES, R.H. 1980. Managing Our Way to Economic Decline. *Harvard Business Review*, No. 4 (July-August).

ADLER, P. 1983. *Rethinking the Skill Requirements of New Technologies*. HBS 84-27.

–. 1984. New Technologies, New Skills. Draft paper for *Harvard Business Review*.

AYRES, R.U.; MILLER, S.M. 1982. Robotics Realities: Near-Term Prospects and Problems. *Annals (AAPSS)*, Vol. 470, pp. 28-55.

BELL, D. 1973. *The Coming of Post-Industrial Society*. New York, Basic Books.

BLUMBERG, M.; GERWIN, D. 1984. Coping with Advanced Manufacturing Technology. *Journal of Occupational Behaviour*, Vol. 5.

BORTNICK, J. 1983. Information Technology and the Developing World: Opportunities and Obstacles. *The Information Society*, Vol. 2, pp. 157-69.

BORTNICK, J.; GILROY, A.; SIDDALL, D. 1984. *A Glossary of Selected Telecommunications Terms*. Washington, D.C., Library of Congress.

BRAVERMAN, H. 1974. *Labor and Monopoly Capital*. New York and London, Monthly Review Press.

BRIGHT, J.R. 1958. Does Automation Raise Skill Requirements? *Harvard Business Review*, No. 4 (July-August).

BRILLOUIN, L. 1956. *Science and Information Theory*. New York, Academic Press.

CHI, S. 1982. Advances in Computer Mass Storage Technology. *Computer*, Vol. 15, No. 5, pp. 60-74.

COLMAN, D.; NIXSON, F. 1978. *Economics of Change in Less Developed Countries*. New York, John Wiley.

CREMIEUX-BRILHAC, J.L. 1980. Les banques de données et les transferts de culture. In: *Actes du Colloque international Informatique et Société*, Vol. 4, pp. 101-10. Paris, La Documentation française.

DE SOLLA PRICE, D.J. 1978. India as a Small, Highly Developed Scientific Nation. In: K.D. Sharma and M.A. Qureski (eds.), *Science, Technology and Development: Essays in Honour of Prof. A. Rahman*. New Delhi, Sterling Publishers.

DERTOUZOS, L.M.; JOEL, M. (eds.). 1981. *The Computer Age: A Twenty-Year Review*. Cambridge, Mass., MIT Press.

DRAPER, R. 1985. The Golden Arm. *New York Review of Books* (24 October), pp. 46-52.

EDQUIST, C.; JACOBSSON, S. 1985. India and the Republic of Korea: Comparative Experiences in Engineering Industries. *ATAS Bulletin*, No. 2, pp. 39-42.

EISENSTEIN, E.L. 1979. *The Printing Press as an Agent of Change*. Cambridge, Cambridge University Press.

ERBER, F.S. 1985. Microelectronics Policy in Brazil. *ATAS Bulletin*, No. 2, pp. 119-24.

ERES, B.K. 1983. Information Technology: Status, Trends and Implications. *Electronic Publishing Review*, Vol.3, No. 3, pp. 223-44; Vol. 3, No. 4, pp. 303-18.

ERES, B.K.; RAZ, B. 1986. Capacity Development for Informatics and its Socioeconomic Implications. Paper presented at the meeting of the International Federation of Information Processing Societies, Dublin, Ireland, September 1986.

EVANS, C. 1979. *The Mighty Micro*. London, Victor Gollancz.

–. 1981. *The Making of the Micro*. London, Victor Gollancz.

FADEM, J. 1984. Automation and Work Design in the United States. In: F. Butera; J.E. Thurman (eds.), *Automation and Work Design*, pp. 647-722. A study prepared for the International Labour Office, Geneva, North Holland Publishing Company.
FISCHER, D. 1981. Lessons from Major Accidents: A Comparison of the *Three Mile Island* Nuclear Core Overheat and the *North Sea Platform Bravo* Blowout. Executive Report No. 6, International Institute for Applied Systems Analysis (IIASA), Laxenberg, Austria.
FREEMAN, C.; CLARK, J; SOETE, L. 1982. *Unemployment and Technical Innovation*, London, Francis Pinter.
Futuribles. 1984 (September). Quelle informatique pour quel développement? Paris.
GASSMAN, H.P. (ed.). 1981. *Information Computer and Communications Policies for the 80's*. Amsterdam, North Holland Publishing.
GELDENS, M. 1984. Toward Fuller Employment: We have been Here Before. Interview with *The Economist* (28 July), pp. 19-22.
GEROLA, H.; GOMORY, R.E. 1984. Computers in Science and Technology: Early Indications. *Science*, Vol. 225, No. 4657, pp. 11-18.
GRUNWALD, J.; FLAMM, K. 1985. *The Global Factory: Foreign Assembly in International Trade*. Washington, D.C., The Brookings Institution.
HIRSCHHORN, L. 1984. *Beyond Mechanism: Work and Technology in a Postindustrial Age*. Cambridge, Mass., MIT Press.
HIRSCHMAN, A.O. 1958. *The Strategy of Economic Development*. New Haven, Yale University Press.
HUDSON, H. 1983. International Satellite Policy Formulation: Issues in Participatory Planning. Paper presented at the Eleventh Annual Telecommunications Policy Research Conference, Annapolis, Maryland, 24-7 April.
JACOBSSON, S.; SIGURDSON, J.(eds.). 1983. *Technological Trends and Challenges in Electronics; Dominance of the Industrialized World and Responses in the Third World*. Research Policy Institute, University of Lund, Sweden, AV-Centralen, Lund.
JAIKUMAR, R. 1984. Flexible Manufacturing Systems: A Managerial Perspective. Working Paper, Division of Research, Harvard Business School, Boston, Mass.
JUNNE, G. 1984a. The Impact of Automation in Industrial Countries on Manufacturing Exports from Newly Industrialized Countries. Workshop on Europe and the New International Division of Labor. Salzburg, Austria and Universiteit van Amsterdam.
–. 1984b. New Technologies: A Threat to Developing Countries' Exports. Contribution to the Seminario 'Revolucion Tecnologia y Empleco'. Mexico City and Universiteit van Amsterdam.
KAPLINSKI, R. 1984. Micro-electronics and the Third World. *Radical Science Journal*, pp. 37-51.
KELLEY, M.R. 1984. New York Technology and its Work Force Implications: Union and Management Approaches. In: H. Brooks et al., *Technology and the Need for New Labor Relations*. Kennedy School of Government, Discussion Paper Series, No. 129D.
–. 1985. Implications of Programmable Machines: *ATAS Bulletin*, No. 2, pp. 95-9.
KING, A. 1982. For Better and For Worse. *Information Society*, pp. 35-57.
KRUPP, H. 1982. Economic and Societal Consequences of Information. *Information and Innovation*, pp. 27-47.
LEBEAU, A. 1972. Plaidoyer pour l'espace. *La Recherche*, Paris, Seuil.
LEFT, H. 1984. Externalities, Information Costs, and Social Benefit — Cost Analysis for Economic Development: An Example from Telecommunications. *Economic Development and Cultural Change*, pp. 255-76.
LEONTIEV, W.; DUCHIN, F. 1985. *The Future Impact of Automation on the Workplace*. New York, Oxford University Press.
LUSSATO, B. 1983. Note sur l'impact des progrès de l'informatique sur les programmes de l'Unesco. Paris.
MACCOBY, M. 1984. Technology, Organization and Leadership. Keynote Address, First International Symposium on Human Factors in Organizational Design and Management, Honolulu.
MATTELART, A.; SCHMUCLER, H. 1985. *L'ordinateur et le Tiers Monde — L'Amérique latine à l'heure des choix télématiques*. Paris, Maspero.
MCHALE, J. 1976. *The Changing Information Environment*. Colorado, Westview Press.

MELODY, W. 1983. Development of the New Communication and Information Industries: Impact on Social Structures. Paper presented at the Symposium on the Cultural, Social and Economic Impact of Communication Technology, Unesco and Instituto della Enciclopedia Italiana. 12-16 December, 1983. Rome.

MELO TELES de, J. 1985. *Pela Valorizaçao da Inteligencia*, Cadernos, Editora Universidade de Brasilia.

MENOU, J.M. 1983. Cultural Barriers to the International Transfer of Information. *Information Processing and Management*, Vol. 19, No. 3, pp. 121-9.

MOLL, P. 1983. Should the Third World have Information Technology? *IFLA Journal*, Vol. 4, pp. 296-308.

MOREHOUSE, tr.; CHOPRA, R. 1983. *Chicken and Egg: Electronics and Social Change in India*. Research Policy Institute, Technology and Culture Series, No. 10. Lund.

NICHOLAS, I.; WARNER, M.; HATTMAN, G.; SORGE, A. Automating the Shop Floor: Applications of CNC in Manufacturing in Great Britain and West Germany. *Journal of General Management*, Vol. 8, No. 3, pp. 26-38.

NILSEN, S.E. 1979. The Use of Computer Technology in Some Developing Countries. *International Social Science Journal*, Vol. 3, pp. 513-29.

NOBLE, D.F. 1984. *Forces of Production: A Social History of Industrial Automation*. New York, Alfred A. Knopf.

NORA, S.; MINC, A. 1981. *The Computerization of Society: A Report to the President of France*. Cambridge, Mass., MIT Press.

OECD. 1980. *Technical Change and Economic Policy*. Paris.

–. 1982. *Micro-Electronics, Robotics and Jobs*, ICCP, Vol. 7. Paris.

–. 1985. *Transborder Data Flows*, Amsterdam-New York, North Holland.

OETTINGER, A.G. 1984. The Information Evolution: Building Blocks and Bursting Bundles. Program on Information Resources Policy, Center for Information Policy Research, Harvard University. Cambridge, Mass.

–. 1985. Automation Technologies: Functional Descriptions. *ATAS Bulletin*, No. 2, pp. 22-8.

OTA. 1984. *Computerized Manufacturing Automation: Employment, Education and the Workplace*. Office of Technology Assessment, United States Congress, Washington, D.C.

PIORE, M.J.; SABEL, C.F. 1984. *The Second Industrial Divide: Possibilities for Prosperity*. New York, Basic Books.

PITRODU, S. 1976. State of Telecommunications in Developing Countries — An Overview. *IEEE Transactions on Communications*, Vol. 24, No. 7, pp. 676-83.

PORAT, M. 1977. *The Information Economy*. National Science Foundation.

RADA, F.J. 1980. Microelectronics and Information Technology: A Challenge for Research in the Social Sciences. *Social Science Information*, Vol. 19, pp. 435-65.

–. 1982. Technology and the North-South Division of Labour. *IDS-Bulletin*, Vol. 13, No. 2, pp. 5-13.

–. 1983. *La Microelectronica, la Tecnologia de Informacion y sus Efectos en los Paises en Via de Desarrollo*. Mexico, El Colegio de Mexico.

RAZ, B. 1983. Education and Infrastructure Consideration in Developing Countries. *Africon' 83. African Electrical Technology Conference* F1.11 -F.1.1.3.

–. 1984. On the Issue of Technology Transfer to Developing Countries. *Global Perspective of the Strategy of Science and Technology for Development*, pp. 1-10.

RICE, E.D.; PARKER, B.E. 1979. Telecommunications Alternatives for Developing Countries. *Journal of Communication*, Vol. 129, No. 4, pp. 125-36.

SABEL, C.F. 1982. *Work and Politics: The Division of Labor in Industry*. Cambridge Studies in Modern Political Economies. New York, Cambridge University Press.

SAUNDERS, R.; WARFORD, J.; BJORN, W. 1983. *Telecommunications and Economic Development*. Baltimore, Johns Hopkins.

SCHNEIDER, L. 1984. Technology Bargaining in Norway. In: H. Brooks et al., *Technology and the Need for New Labor Relations*. Kennedy School of Government, Discussion Paper Series, No. 129D.

SCHUMAN, G. 1984. The Macro- and Microeconomic Social Impact of Advanced Computer Technology. *Futures*, pp. 260-85.

Servan-Schreiber, J.-J. 1981. *Le défi mondial*. Paris, Hachette.
Shaiken, H. 1985. *Work Transformed — Automation and Labour in the Computer Age*. New York, Holt, Reinehart and Winston.
Shannon, C.E.; Weaver, W. 1949. *The Mathematical Theory of Communication*. Urbana, Illinois, University of Illinois Press.
Skinner, W. 1983. Wanted: Managers for the Factory of the Future. *Annals* (AAPSS), Vol. 470, pp. 102-14.
Smith, M.J. 1980. VDT's: A Preliminary Health Risk Evaluation. National Institute of Safety and Health (NIOSH), Cincinnati, Ohio.
Spiegel-Rosing, I.; De Solla Price, D.J. (eds.). 1977. *Science, Technology and Society: A Cross-Disciplinary Perspective*. London, Sage Publications.
Stonier, T. 1983. *The Wealth of Information: A Profile of the Post-Industrial Economy*. London, Methuen.
Sultz, J. 1984. *Informatique et société: Quelques réflexions à partir du Tiers Monde*, Thèse de doctorat Paris I-Sorbonne, Iedes (mimeographed).
Swedish Work Environment Act, January 1983. Stockholm.
Swedish Act on Codetermination at Work, January 1985. Stockholm.
Telesis Consultancy Group Inc. 1982. *A Review of Industrial Policy*. National Economic and Social Council, Republic of Ireland, Report No. 64. Dublin.
Toda, M. 1984. Employment Effects of Microelectronics Technology in Japan. International Institute for Advanced Study of Social Information Science, Namazu, Japan, Fujitsu Ltd.
Unesco. 1982. *The Use of Satellite Communication for Information Transfer*. Paris, The International Institute of Communications.
Van Der Loo, H.; Slao, P. 1982. Information Technology. *Telecommunications Policy*, pp. 100-10.
Walton, R.E. 1985. From Control to Commitment in the Workplace. *Harvard Business Review*, No. 2 (March-April).
Wied, A. 1982. Microelectronics: Implications and Strategies for the Third World. *Third World Quarterly*, Vol. 4, No. 4, pp. 677-97.
Williams, Sir Bruce. 1983a. Technology Policy and Employment. *Higher Education* (Amsterdam), Vol. 12, pp. 431-42.
–. 1983b. Long Term Trends of Working Time and the Goal of Full Employment. OECD, Paris.
Zhang, X. 1985. Microelectronics Policy in China. *ATAS Bulletin*, no. 2, pp. 127-9.

3

Biotechnologies in farming and food systems

Introduction

Agriculture is a high priority for almost all developing countries: they face an overwhelming need to increase food production in order to feed their growing populations and — since most of their economies are still dominated by agriculture — to produce a surplus of cash crops for export.

For centuries agricultural production was limited by the availability of land and the labour to cultivate it, and by the nature of the climate and soil. The introduction of machinery, more efficient irrigation systems and increasingly sophisticated plant breeding techniques permitted a very substantial increase in productivity within the overall constraints mentioned. The latest technological developments, the biotechnologies, promise improvements in agriculture of a completely different order of magnitude, releasing farmers from many of the traditional constraints and hazards.

Biotechnologies are an array of tools derived from research in cellular and molecular biology that could be applied in any industry that involves micro-organisms, as well as animal and plant cells (Sasson, 1984). The techniques promise to transform agriculture, and they also have importance for other carbon-based industries such as energy, chemicals, pharmaceuticals and waste management. The potential impact is enormous, because research in the biological sciences is making very rapid advances and the results not only affect a range of sectors but also promote greater linkage between them. A research success in fermenting energy from agricultural wastes, for example, would affect both the energy sector and the economics of food production.

For developing countries — still predominantly agricultural — the results of the 'biorevolution' could be highly beneficial. On the other hand, the availability of the new techniques raises issues that are more disquieting. Agriculture will certainly be more productive, but will this increased productivity benefit those who are too poor to purchase the output? Scientific advance is very rapid, but will those producers who are among the world's poorest be able to take advantage of the latest techniques? Will developing nations be able to devise ways of ensuring that they do not become hopelessly dependent upon technologies imported at exorbitant prices, but instead adapt biotechnology to their specific needs at reasonable cost?

Any productive force of the power and magnitude of biotechnology will necessarily have tremendous influences and diverse impacts on technical, economic and social systems. There will be winners and losers, but no group is necessarily 'fated' to be a loser, since that depends on the prosecution and outcomes of the strategies a group, country, or group of countries adopts to ensure that they secure their rightful share of the benefits. Choosing not to participate in the 'biorevolution' is not a viable strategy because the impacts cannot be avoided, and even the most isolated societies will be affected.

The beauty of biotechnology is that it has many levels of sophistication, ranging from capital-intensive to labour-intensive, from the ultra-high technology of the biology department of a major American university with its multimillion dollar budget to family-size tissue culture laboratories operating in the Socialist Republic of Viet Nam. Many of the ultra-sophisticated modern techniques can be reduced to relatively low-cost operations without losing all possibility of success. Clearly, American research laboratories and family-size tissue culture laboratories are aiming at very different goals, but this is precisely the point: less developed countries must aim at achievable targets appropriate to their own circumstances. It would be a certain mistake simply to attempt to imitate the research and activities of the developed countries.

Traditional farming systems: problems and strategies

'TRADITIONAL' AND 'MODERN' SYSTEMS

An agricultural or farming system as a bioeconomic activity consists of a farm enterprise or business in which the farmer manipulates various environmental factors and manages resources and inputs in order to achieve with varying degrees of success objectives for producing food, feed, fibre, drugs, etc., required by man. The farming system may entail growing one or more species of crops, or else rearing one or more species of animals, perhaps in combination with one more species of crops. The different farming systems that characterize various cultural landscapes differ from each other in relation to the various prevailing factors of production: physico-chemical (climate, soils, nutrients, water), biological (crops, animals, weeds, pests, pathogens), technological (tools, machines, practices), managerial (knowledge, experience, decision-making, skill) and socioeconomic (markets, labour, religion, customs, credit, personal preferences) (Okigbo, 1982).

There currently exist side-by-side in both developed and developing countries agricultural production systems that for convenience may be classified as 'traditional', 'transitional' and 'modern' or commercial farming systems. The traditional farming system is one in equilibrium with the times, and is the culmination of several millennia of trial and error and interaction between men, environment and organisms (Shultz, 1964). Modern farming systems developed out of changes in traditional systems over the last century as a result of applications of science and technology to the improvement of agricultural production. This classification does not imply that traditional and transitional systems are static, however. For example, farming systems in tropical Africa have changed significantly over the centuries as new crops were introduced from Asia and America and, more recently, with improved transport and communications, wider markets and mechanization of production.

Traditional and transitional farming systems have characteristics very different from 'modern' farming systems: typically they are small (under 5 ha) subsistence units operated by family members and cultivated manually with little use of any commercial inputs, improved seeds or irrigation. There are labour shortages at peak times, followed by long slack periods. In many cases, slash-and-burn clearance and fallowing are used to maintain soil fertility. The advantages of this system are the use of intercropping which assures reasonable yields with a minimum of risk — in its purest form the traditional system is designed to maximize risk aversion (Ruthenberg, 1980; Okigbo, 1982).

CONSTRAINTS TO INCREASED PRODUCTIVITY

In the past, traditional agricultural production systems were ecologically sound and constituted efficient use of resources, given the objectives and the ecological and socioeconomic conditions under which they were evolved in different ecological zones. They have become increasingly outmoded, however, when faced with rapid population growth and pressures of 'modernization'. Constraints to their improvement are the result of interrelated physical, biological and socioeconomic factors (Flinn et al., 1974; Okigbo, 1982, 1984).

The physical constraints arise in part from unfavourable climatic conditions, including unpredictable periods of drought, floods and environmental stresses, with rainfall that is unreliable in onset, duration and intensity. Rainfall may be less effective in sandy soils and on steep slopes, while tropical rainstorms cause problems of erosion and cloudiness reduces photosynthetic efficiency. In addition, most soils of the humid and subhumid tropics have very low inherent fertility (except on hydromorphic and young volcanic soils), high rates of decomposition and low levels of organic matter, very low cation exchange capacity (CEC) and thus also less active colloidal complex; they are often intensely weathered, sandy and low in clay, with very high acidity and sometimes high surface temperatures, too high for some crops and biological processes such as nitrogen fixation. Natural problems tend to be exacerbated by farming methods: multiple nutrient deficiencies and toxicities develop under continuous cultivation, for example, and serious salinity problems result from poor irrigation management.

The biological constraints are related to the high incidence of disease, pests and weeds owing to environments that favour causal organisms. Radical environmental changes have been brought about by human activities that have adverse effects on ecological equilibrium.

Socioeconomic constraints include such factors as the small size of farms, drastically reduced by population pressure, and unfavourable land tenure systems which often result in fragmentation of holdings. Production is further hampered by low incomes and lack of credit, coupled with the high cost and extreme scarcity of inputs, while poor marketing facilities and transportation make it difficult to export any surplus. Illiteracy and superstition sometimes hinder the adoption of new techniques, a process which in any case tends to happen in piecemeal fashion.

The overall result is that there is a wide gap between yields on farmers' fields and research station yields. Technology by itself is not a panacea, as experiences in the 'Green Revolution' have shown; the new techniques must be socially and culturally acceptable as well as economically and technically viable for the particular setting, and farmers must be both aware of their existence and able to understand, use and adapt them to their needs and circumstances. There have been practical difficulties, too, where all the elements in a new package are not available in time and or in sufficient quantities (McLoughlin, 1970). Improvement in this situation requires political commitment, institutional changes and a favourable socioeconomic climate which would ensure the institutionalization of the essential factors that favour and accelerate agricultural development (Mosher, 1970; see Table 1).

Table 1. Essential factors and accelerators of agricultural development.

Essential factors

- Continuous and systematic research that generates new agricultural technology
- Adequate incentives for farmers' agricultural services operations, etc.
- Effective transportation and communication system that reaches most farms
- Local availability at reasonable prices of farm equipment and inputs, either manufactured locally or imported
- Effective marketing system for agricultural produce

Accelerators

- Availability of production inputs
- Opportunity for group action by farmers in cooperatives and other special organizations
- Means for improving and expanding land under cultivation
- Adequate educational and training facilities for agricultural technicians and experts for all supporting services
- Mechanism for planning and directing agricultural development programmes as integral component of overall economic development

Source: Mosher, 1970.

ALTERNATIVE STRATEGIES FOR IMPROVING PRODUCTION

Even if it were possible, traditional farming methods clearly should not be abandoned completely, nor should the latest techniques be applied blindly in all settings: evaluation of both old and new technologies is essential, and their most positive features must be identified and incorporated into strategies for the future.

Table 2. Desirable characteristics of traditional and modern farming systems.

Traditional systems

Diversification of production through temporal and spatial improvement of multiple cropping patterns that ensure satisfaction of the farmer's subsistence and increasing cash requirements, while maintaining stability of production and reducing risks.

Integration of crop and animal production systems in addition to development of farming systems that involve components of improved forestry and silvopastoral systems as circumstances permit.

Utilization of nutrient cycling and biological nitrogen fixation potentials of plants wherever possible in order to reduce the need for costly fertilizers.

Cropping systems which make as much use as possible of indigenous and under-utilized crop plants.

Development of improved cropping patterns, grazing systems and technologies which ensure that the soil is kept adequately protected from erosion and degradation.

Integrated watershed development including the development and utilization of relatively more fertile valley bottoms and hydromorphic soils.

Modern systems

Mechanization and appropriate technology to minimize drudgery in farm work while significantly increasing productivity.

Integrated pest management to reduce losses in the field and in storage.

Techniques and methods to increase the efficiency of those fertilizers which cannot yet be replaced by biological nitrogen fixation.

Intensification of production and increased productivity per unit area of land in order to curtail drastically the reliance on expansion of area under cultivation as the main strategy for increasing production.

Increased use of irrigation and water harvesting in semi-arid and arid areas, with measures taken to ensure adequate drainage and to minimize salinization.

Methods to reduce tillage, even to zero level.

Utilization of the available old and new results of conventional agricultural genetics.

Judicious use of agricultural chemicals.

Source: Okigbo, 1984.

The farming systems approach (FSR) has proved effective as a means of analysing techniques in practice. This involves an initial survey and diagnostic stage, followed by research-station evaluation of the data and design of suitable technologies which are then tested on the farmers' fields, the process and extent of their adoption are monitored so that necessary modifications can be made, and the results are fed back to research-station scientists. FSR facilitates understanding of farming methods, identifying the scientifically sound and otherwise desirable aspects of both traditional and modern production systems, which can then be improved or repaired by modifying them or designing new ones. Some of these desirable characteristics, which should be taken into consideration in developing priorities and strategies for increased food and agricultural production, are summarized in Table 2.

In the development of new technologies and associated production systems, three options are available:

— modification of the farmer's agroecosystem

— synthesis of new systems based on agroecosystems design and management principles

— mimicking of analogous natural ecosystems.

The new systems that result from modifications of traditional farming methods do not involve drastic changes which may make them difficult for farmers to adopt — the major disadvantage is that modification may bring about only slight changes which may not be very attractive to farmers. Synthesis of new systems has the major disadvantage that at present design principles and environmental information are not available, and the agroecosystem management plan produced may require too radical a change from the farmer's management plan, thus significantly reducing chances of adoption. There is also the problem that the new system involves the constant evaluation and improvement of design principles as better principles and technologies become available.

The mimicking of analogous ecosystems requires very little analysis of ecological factors since it is usually assumed that the natural system is well adapted to prevailing environmental conditions. For example, in cropping systems similar to the natural ecosystems, weed competition is minimized because all the elements in the system are structurally and functionally compatible. The main disadvantage of the approach is that natural ecosystems are closed systems with little biomass for export, so that similar agroecosystems are also very likely to be low yielding. It has been observed, however, that in situations where a few species of tree crops similar in floristic composition and structure are grown in polyculture (such as palms, bananas and cocoyams), some reasonably good yields may be obtained.

It is obvious that the most feasible option, especially where time is a constraint in the race against population growth and food production, is the modification of the farmer's production system. FSR usually facilitates this and calls for a strategy of developing sequentially new technologies and production systems through integration of traditional and conventional and or emerging technologies wherever appropriate. This approach has not so far eliminated the wide yield gap between experimental condition and the farmer's field, for several reasons. Firstly, FSR is but a recent development in the International Agricultural Research Centers network (see p. 144) and a few national programmes; it requires multidisciplinary research teams and more effective cooperation and linkage of research, training and extension than many individual developing countries are able to accomplish. Limitations in the scientific manpower capabilities, technological infrastructure and support, financial resources, and deficiencies in rural infrastructure and policies make it difficult for promising new production systems and component technologies to be adopted by farmers or to be more finely tuned to the local circumstances of most traditional farmers. If these difficulties have been encountered in the implementation of relatively well-established technologies, it is clear that there will be similar problems in introducing the results of the most recent biotechnological research.

Major plant biotechnologies and their applications

Agriculture has already benefited enormously from scientific advances in recent decades, in particular from the more sophisticated plant breeding techniques which have created high yield crops better adapted to particular growing conditions, and from improvements in agricultural chemicals (fertilizers, herbicides, pest control) (see Table 3). The biotechnologies promise benefits on a far greater scale, in that they not only vastly improve the efficiency of conventional processes, but they permit the development of completely different techniques that bypass fundamental problems and, in the longer run, will create totally new crops and products.

Research in cellular and molecular biology has provided a much better understanding of plant behaviour, as well as the means to modify that behaviour through the manipulation of genes and cells so as to satisfy increasingly specific requirements. The technologies can be grouped under the two general headings of tissue culture and genetic engineering. The first operates at or above the level of the cell (with its components — chloroplasts, mitochondria, etc.), and involves growing cells, tissues and organs in controlled conditions. The second concerns the manipulation of the genes that determine cell (and therefore plant) characteristics, which means working at the level of DNA: the isolation of genes, their recombination and expression in new forms and transfer into appropriate cells.

Table 3. Average yields per acre in 1930 and 1975 in the United States of America.

Crop	1930	1975	unit	Percentage increase
Grain sorghum	10.8	49.0	bushels	115
Corn	20.4	86.2	bushels	320
Potatoes	65.9	251.0	cwt	311
Peanuts	659.4	2565.0	pounds	295
Cotton	157.0	453.0	pounds	188
Tomatoes	61.0	166.0	cwt	172
Rice	21.0	45.6	cwt	117
Wheat	14.2	30.16	bushels	115
Soybeans	13.4	28.4	bushels	112
Barley	24.0	44.0	bushels	85
Sugarbeets	11.9	19.3	tons	62
Oats	32.2	48.1	bushels	50
Alfalfa	1.95	2.87	tons	42

Source: Sprague et al., 1980: 17.

With these technologies, it is possible to create — more rapidly and accurately than before — new plant varieties that have improved characteristics (e.g. higher yields, tolerance to conditions such as salinity or waterlogging, resistance to specific herbicides and diseases), and that can be propagated more successfully. Even certain sexually incompatible plants may now be hybridized, and the potential range of new varieties is immense. Problems of disease and weed control are beginning to be tackled genetically rather than with chemicals, and work is underway to produce nitrogen-fixing plants.

Much research still remains to be done (certain phenomena are as yet poorly understood, and techniques, e.g. of handling DNA, are evolving constantly), and many of the applications are still a long way from successful implementation, but the potential advantages and disadvantages can already be identified, at least in outline.

TECHNIQUES

Plant tissue culture

At the heart of the newly developing biotechnologies are the techniques of isolating cells, tissues or organs from plants and growing them under controlled conditions (*in vitro*). (The term 'tissue culture' is used here for brevity, to cover cells, tissues and organs). A considerable range of techniques is already available, varying greatly in sophistication and in the time-horizon required to produce useful results. (A fuller technical discussion of the techniques involved can be found in Sasson, 1984, 1986a, b.)

At the simplest level, vegetative propagation from cuttings has long been practised, with ever increasing efficiency. The method is cheap, rapid and effective for many species (in particular most major fruit and flower species, and some key vegetables), although not all (Conger, 1983). *In vitro* propagation techniques include meristem cultures (which make use of the small mass of undifferentiated cells at the tip of the stem that grows constantly and that generates the plant organs), embryo production by somatic embryogenesis, and shoot production by organogenesis.

The advantages of *in vitro* propagation are:

— production of large numbers of plants or clones in a short time and using small confined facilities

— propagation of materials in an environment free of viruses and other pathogens, and under optimum conditions

— ability to propagate plant species which are difficult to propagate vegetatively once they flower, or to propagate species where *in vitro* culture techniques are commercially superior to other conventional methods of propagation (e.g. for tropical plants with high levels of tannin and phenols)

— ability to supply plants on a year-round rather than on a seasonal basis

— maintenance of heterozygozity and the cloning of superior individuals from both qualitative and quantitative viewpoints. Fruit trees, oil, date and coco-palm trees, species for lumber and pulp are good candidates for such *in vitro* cloning and, in fact, outstanding results have been obtained with the palm species, and with apple, cherry, rhododendron, eucalyptus, aspen and pine trees (Bonga and Durzan, 1982; Vidalie, 1983).

There are nevertheless some disadvantages, such as the risk of curtailing genetic diversity, the long time-periods involved in culture establishment of some species (although multiplication can be rapid later on), the relatively high cost of *in vitro* propagation (compared with propagation by root cuttings), and the need for qualified staff and laboratory equipment. Moreover, *in vitro* propagation remains problematic for many crops, and future progress on cereals and grain legumes will require intensified crop-specific work and an understanding of the biochemical basis for plant regeneration and its control.

Since 1968 there has been a dramatic increase in the number of plant species propagated *in vitro* (30 species by 1968, and more than 300 species belonging to more than 40 families by 1978) and there is the prospect of a further tenfold increase by 1988.

Somaclonal variation

Early efforts in tissue culture attempted to produce exact clones, but it was soon discovered that the culture process itself caused variations (termed 'somaclonal' variation) in differing degrees depending on the species; somaclonal variation also occurs naturally in many species, at a low rate of frequency (one in every million cells). This instability has both positive and negative aspects. It is a problem that must be controlled in some circumstances — where it is essential to maintain varietal purity in the propagating material, especially in the case of clonal propagation of elite genotypes. On the other hand, the variation can be exploited to increase genetic diversity and, in three to ten years, create new plant varieties (Table 4). Such a technique will then allow the release of new varieties in approximately half the time required by conventional breeding, including the time of crossing, back-crossing, selfing and field trial selections.

Table 4. Timetable for new variety development by somaclonal variation of elite cultivars.

Crop species	Conventional breeding (years)	Somaclonal variation (years)
Tomato plant	7-8	3-4
Sugar-beet	14-15	7-8
Sugar-cane	14	7
Coffee-tree	15-20	7-10

Source: Sondhal et al., 1984: 17.

Somaclonal variants include resistance to herbicides and to diseases: sugar-cane, for example, has developed resistance to eyespot disease, Fiji virus, downey mildew and smut; potatoes to late and early blight; maize to Southern maize blight; and rape to vitricular disease. Somaclonal variation may also provide genotypes suitable for tropical environments, i.e. able to tolerate heat or acidic soils containing harmful concentrations of aluminium and manganese. In crop species where the creation of genetic variability through sexual reproduction is difficult, somaclonal variation along with mutagenesis is a promising way to generate variability for crop improvement. In sexually reproducing species where an *in vitro* screening procedure is available or can be designed, somaclonal variation can be useful.

Australian researchers have found that stable variation occurs in wheat, in multigenetic traits such as height and maturation date, as well as in single-gene traits (Scowcroft and Larkin, 1982). They are now screening cells in culture in search of traits that would be useful in no-tillage farming, which is increasingly being practised as a means of conserving soil.

Longer-term techniques

In vitro propagation is not the only method for crop improvement; rather it is an additional tool for breeding and other related research programmes. That is why, once a useful crop has been established by simple clonal propagation, it could be further improved by using both conventional breeding techniques and long-term tissue culture techniques, so that a collaborative effort between plant breeders, phytopathologists, physiologists, food scientists and biotechnologists is imperative for the final improvement of a particular species.

These longer-term tissue culture techniques include anther and pollen grain culture, embryo culture, *in vitro* fertilization of cultivated ovaries and somatic hybridization. Anther culture was developed in India in the early 1960s and has since been experimented with and refined in several countries. It allows the fixation of desired characteristics much more quickly, thereby reducing breeding time significantly: for example, breeding time for a new barley variety was reduced from twelve to five years through the use of the so-called *bulbosum* method (Kasha and Reinbergs cited in Evans et al., 1983). Already an International Rice Research Institute anther-cultured, cold-tolerant rice is being field tested in Korea and the technique has been effectively exploited in China for producing new rice and wheat varieties. A long list of other crops that have benefited from progress in this area includes vegetables such as cabbage, aubergine, potato, tomato; fodder and industrial species (rapeseed, tobacco); and ornamentals. Anther-culture also facilitates the crossing of domestic plants with their wild relatives, thereby incorporating new material into the gene pool.

Embryo culture involves rescuing immature embryos from defective seeds (which result from various mechanisms of sexual incompatibility in certain hybrids), and then growing them on a suitable culture medium to the stage of viable seedling. Recovery of an F^1 hybrid in this way must be followed by several cycles of back-crossing and selfing to select the desired traits, and the whole process through to production of the final commercial variety may require five to fifteen years. The techniques have been successful with orchids, tomato, cotton, beans, barley, banana and soft coconut.

Somatic hybridization involves the fusion of the protoplasts of two somatic plant cells and is a means of extending the range of possible hybrids to produce traits that had not been obtainable by other methods (see Sasson, 1984:116ff. on potato cultivars resistant to disease) and to overcome barriers of sexual incompatibility between sexually distinct species (e.g. the 'pomato' created by Danish and German scientists in 1978), which would allow traits developed genetically in one plant to be transferred to another, less susceptible to genetic improvement, in a single operation rather than over many years of conventional breeding.

The same technology could also be used to explore limited gene transfer for the disease resistance and salinity tolerance in rice, as well as to produce male sterile and triazine-resistant lines. Applications for crop improvement also clearly exist for protoplast fusion procedures in wheat, barley, maize, in several legume species once the technique of regeneration of plants from the protoplasts of these species is mastered. The fusion of protoplasts of rice and *Azolla* cells (a fern which forms nitrogen-fixing symbiotic associations with a cyanobacterium, *Anabaena*) is being studied by the researchers of the University of Nottingham and of the International Rice Research Institute (Swaminathan, 1982).

In vitro techniques now allow efficient recovery of mutant cell lines because plant cells can be manipulated in a similar fashion to micro-organisms. The advantages of *in vitro* mutant isolation over conventional field screening include the application of greater selection intensities and the substantial reduction in time, expense and resource use. The plant tissue culture specialist prefers to select the improved traits (such as disease resistance or herbicide and salinity tolerance) at the cellular level because of the size of the cell population (1 ml of a plant cell culture can contain about 1 million cells). After randomly recovering plants from the cell cultures, selection can also be made at the plant level, in the greenhouse or in the field.

Physical or chemical mutagens can be used to mutate the cells and they will increase the variability by 100 to 1,000 times and may favour the appearance of the desirable character. The application of this technique has permitted the recovery of maize plants mutant lines with increased threonine content (33 to 59% higher threonine), tobacco plants with more lysine (10 to 15%), petunia plants resistant to mercuric chloride, tobacco plants resistant to salt and herbicides such as picloram and paraquat. At the International Rice Research Institute, in the Philippines, the researchers are trying to isolate rice mutant cell lines which could tolerate high concentrations of salts and aluminium, as well as varieties with higher protein and lysine contents in their seeds (Swaminathan, 1982). Cell and tissue culture may lead to the isolation of superior plants having a higher photosynthetic efficiency and productivity.

Cell or tissue lines could also be helpful for the selection of plants resistant to pathogens. Such a technique was used to detect plant varieties which are resistant to toxins produced by fungi and bacteria. For instance, Gengenbach and Green have isolated a maize cell line resistant to the toxin of the pathogenic fungus, *Helminthosporium maydis* (T race), and have regenerated whole plants from these cells. Similarly, cell lines resistant to *Helminthosporium sacchari* toxin have been isolated from sugar-cane tissue cultures.

In vitro selection is, however, limited to crop species for which efficient tissue culture and plant regeneration already exist. Moreover, it can be applied only to agriculturally useful traits which are correlated with cell culture response. Such a correlation is not a frequent event, i.e. the desirable traits observed *in vitro* are not often observed in the whole regenerated plant. A favourable example is the resistance to salinity of *Hordeum vulgare* and *Hordeum jubalum* plants which was identical to that observed in cultivated cell lines; the transmission of traits (size, precocity) selected *in vitro* has been achieved in lettuce, tomato, petunia and sugar-cane (Normand-Plessier, 1984). It is also very important to ascertain the genetic basis of any selected mutant cell line by its progeny analysis. If the new variant does not perform well in the field, it may be necessary to incorporate the selected characteristics into a well-known commercial variety.

The nucleus of a plant cell carries about 95% of the total genetic information, while the remaining 5% is encoded in organelle (mitochondria and chloroplasts) genes. Many desirable traits — such as photosynthetic efficiency, male sterility, resistance to herbicides and to certain diseases — are controlled by organelle genes. The transfer of these traits could be achieved by making enriched preparations of one plant organelle from one species and subsequently incorporating them into the protoplasts of a second species. The success of this technique depends on the isolation and uptake of viable organelles and on the regeneration of complete plants from the transformed protoplasts.

Plastids and mitochondria could also be genetically engineered, but this will require detailed knowledge of the nuclear and organelle genes that, once expressed, lead to mature differentiated forms of these organelles; an understanding of the mechanisms regulating the expression of these genes; and techniques for transforming the nuclear and plastid genomes with foreign DNA. Chromosomes have been isolated from root-tips, microspores and from protoplast cultures of tobacco, onion, peas and lily. The transfer of chromosomes from one plant species to another has been achieved with an efficiency of 1 to 0.001% (Kosuge et al., 1983). Such a technique offers the advantage of transferring genes that are encoded in a particular chromosome which could be isolated by ultracentrifugation, and of polygenic traits linked to a marker.

Genetic engineering in plants

The most critical development for biotechnologies was the discovery that a DNA sequence (gene) inserted into a bacterium induced production of the appropriate protein (Cohen et al., 1973). This opened up the possibilities of gene recombination and transfer, with far-reaching implications for agriculture through the genetic manipulation of micro-organisms, plant and animal traits.

The initial impacts of recombinant DNA agricultural inputs are likely to be on the first of these because of their simplicity and the advanced level of skills now available for manipulating them. Already scientists are proposing to release genetically engineered micro-organisms into the environment deliberately; for example, a genetically engineered bacterium intended to increase plant frost resistance has been developed that is supposed to displace the destructive ice-nucleating bacteria which facilitate ice formation on plants during frosts and thereby prevent frost damage.

The ability to isolate, purify and characterize specific pieces of a plant cell DNA, i.e. the applications of molecular biology and genetics to higher plants, could play an important role in future plant breeding programmes. Moreover, the transfer of genes from one plant to another, i.e. transformation, will provide for the introduction into established cultivars of specific characters controlled by one or a few genes, without disturbing the adopted genetic background. Gene transfer will allow variation in the copy number and in the expression of a given gene, and will permit the insertion of genes from other species.

The successful application of gene transfer and genetic engineering will require gene identification and isolation of those genes or groups of genes which control economically advantageous traits. This process will require success in many areas: gene cloning and sequencing; development of appropriate gene vectors for plant cells; successful gene transfer, gene expression, gene stability and gene transmission through subsequent sexual reproduction (Cocking et al., 1981; Davies, 1981). Although the tools of genetic engineering are available, much better knowledge of plant molecular biology is needed. The recovery of improved crop species through genetic engineering will be the result of a team effort involving at least molecular and cell biologists, plant tissue culture specialists and plant breeders.

It has recently been shown that genes of higher plants are very much like genes of higher animals. Each gene has three regions, all essential for successful funtioning: the beginning is the promoter region which is recognized by the enzyme that triggers the transcription of DNA to RNA; the middle sequence of nucleotides contains the code, i.e. the instructions for producing a specific protein; the end series of nucleotides, or terminator, is a signal to stop the transcription process.

Regulation of plant genes is proving to be more intricate than initially thought. Besides being active or silent, there are some genes that are transcribed into messenger RNA which is, for some reason, never transported to the cytoplasm and to the site of protein synthesis, i.e. this RNA is not translated. Thus, the regulatory mechanisms, in addition to the usual 'on' and 'off' switches determining whether an active gene will actually be expressed, are not yet fully understood. Molecular biologists are still trying to understand the controls of differential expression during plant development to find the DNA sequences that instruct a gene to be 'on' in a leaf, and to be 'off' in the root. Such knowledge and findings are necessary to make gene-splicing techniques more predictable in the engineering of plants.

In tobacco, for example, about 100,000 genes are apparently active at a specific time in the life-cycle of the plant, which means that only about 5% of the DNA contained in the nucleus is used at one time to specify the synthesis of proteins; the function of the other 95% is not known, although regulatory sequences account for some of this DNA. Moreover, about 25,000 genes are 'on' in each organ system of the plant, e.g. in the leaves, stems or flowers. In each organ system there is a unique set of genes that are expressed in it: for instance, the petals and the leaves of tobacco each contain approximately 7,000 specific genes, the ovary and the anther contain approximately 10,000 specific genes (National Research Council, 1984).

Single genes to be used for transformation are being sought, and DNA sequences underlying multigenic traits need to be identified and isolated. A promising approach for both is insertional mutagenesis using transposable elements which can change their position in the chromosomes, and when one of them inserts itself into a dominant single gene, its effect on the plant phenotype can be observed. The gene can be isolated by isolating the transposable element from its position and characterizing the DNA adjacent to the element. In maize this is readily possible with conventional breeding methods, and the use of maize transposable elements for dicotyledonous plants is currently under investigation.

Detailed knowledge is now being developed of the chromosome map of a plant necessary for the identification and subsequent isolation of genes. Many chromosome markers can be generated as a consequence of polymorphisms in DNA sequences throughout the chromosomes. These polymorphisms are easily detected by digestion of DNA with restriction enzymes (endonucleases) and are known as restriction fragment length polymorphisms (RFLP). RFLP mapping is a useful tool for the screening of desirable traits once a RFLP has been located close to the gene of interest.

The establishment of vector systems for plant gene transfer has experienced significant progress in recent years. The most promising candidates up to now are the plasmids of *Agrobacterium tumefaciens*, synthetic plasmids or naked DNA (encapsulated in liposomes), *E.coli* plasmid fused with chloroplast DNA, cauliflower mosaic virus (CMV) or other plant viruses. Very active research projects are underway to develop further methods of transferring DNA into higher plants.

SOME CROP APPLICATIONS

Oil palm

The research progress that has been made since the mid 1970s in oil palm tissue culture is one of the outstanding achievements of biotechnology. The oil palm, which is grown exclusively in tropical developing countries, is important as both a cash crop (a quarter of the production is exported to industrialized countries) and for food. Palm oil is extracted from the pulp and made into edible oils and margarine; cabbage palm oil is extracted from the stone of the fruit and is used in the manufacture of soap, detergents and cosmetics. Output doubled between 1960 and 1980, reaching 4.9 million tons in 1980-1, second only to the soybean (13.4 million tons) among oleagineous plants (Lioret, 1982). World demand for oils continues to increase rapidly so that further development of palm oil production is likely to be very profitable; it would also contribute to an improvement in diet in the developing countries, which lag far behind the industrialized countries in consumption of oils and fats (5.5 kg per person in developing countries as against 20.6 kg in industrialized countries in 1978, according to FAO).

The major problem in oil palm cultivation is that the trees rapidly grow too tall for practical harvesting, so that plantations have to be renewed every 25-30 years, requiring the propagation of millions of plantlets annually. Vegetative propagation is complicated by the fact that the oil palm tree bears male and female flowers which must be cross-pollinated, entailing a high variability in progeny and a lengthy selection process for successful crosses. Research stations in the Ivory Coast and Colombia managed nevertheless to produce much higher-yielding varieties by selective cross-breeding. An additional difficulty for vegetative cloning is that the tree does not grow shoots or put out suckers naturally, nor can small twigs be removed for grafting purposes or for cultivating meristems. The technique ultimately developed involves somatic embryogenesis, i.e. the transformation of tissue calli into embryoids that resemble the embryos resulting from sexual reproduction (for the techniques, see Sasson, 1984:13ff).

The research work was carried out in the Ivory Coast by IRHO and ORSTOM and by Unilever in the United Kingdom. Production of plantlets on a semi-industrial scale began at La Mé in the Ivory Coast in 1981. Cloning material is being provided to other oil palm producers in West Africa and cooperative agreements have been signed with institutions in Colombia, Indonesia and Malaysia for the exchange of plant materials and information on cloning techniques (Lioret, 1982). Other cooperative ventures involve commercial firms in the United Kingdom, Malaysia, India and Japan.

Similar techniques are being applied successfully to other tree crops, in particular the coco-tree (Branton and Blake, 1983).

Other crops

Researchers are constantly seeking to expand the application of biotechnology techniques to crop improvements. One of their goals is the successful development of hybrid rice which would raise yields and extend the conditions under which this crucial food crop could be grown. As already mentioned, work is underway at the International Rice Research Institute using plant tissue and cell culture techniques to develop tolerance of rice strains to salt and aluminium in soils and to improve the protein content. A range of other potential applications is listed in Table 5, of which genetic engineering is the most ambitious since manipulation of the 17 or more genes involved will be extremely challenging (Swaminathan, 1982). The People's Republic of China already has 6 million hectares under hybrid rice strains that are based on cytoplasmic male sterile lines adapted to temperate conditions. A practical problem of hybrid rice will be the prompt distribution of seed to farmers since retained seed cannot be resown.

Table 5. Possible applications of biotechnology research to rice improvement.

Research technique	End result
Tissue and cell culture	
Induction and selection of useful mutants at the cellular level	Salt tolerance Aluminium toxicity tolerance High lysine and high protein Low photorespiration Disease resistance Low oxygen tolerance
Embryo culture	Intra- and interspecific hybridization
Anther and pollen culture	Reducing breeding time
Protoplast fusion	Interspecific and intergeneric hybridization Hybrid rice improvement *Azolla* improvement
Genetic engineering	Incorporation of nitrogen-fixing genes

Source: Swaminathan, 1982: 969.

Another tropical food crop that could benefit from plant tissue culture techniques is the yam. Meristem tip culture, as in the case of the potato, will permit the selection of virus-free plants and so reduce the considerable problems of disease. Micropropagation techniques have been evolved that will overcome the difficulties of large-scale propagation of yams, hampered in the past by the very slow growth of the tubers. Small tubers can now be induced when plantlets derived from stem cuttings are transferred to another culture medium (Gascon, 1982).

OTHER APPLICATIONS AND PROSPECTS

Herbicide resistance

Successful transfer into crop plant species of genes controlling herbicide resistance is an area of intense research interest. One example is herbicide atrazine, that is commonly used in the culture of maize: in the United States of America, for instance, 90% of maize acreage in Illinois is treated with atrazine each year. Maize can tolerate atrazine because it contains an enzyme which breaks down the herbicide, but soybeans (which are often grown in rotation with maize in the Middle West) are susceptible to atrazine. When residues of atrazine remain in the soil, they can dramatically reduce the yield of soybeans planted during the following year. Selecting an atrazine-resistant soybean would be an important and useful achievement for soybean growers, especially in the Corn Belt.

Researchers in the Middle West have studied the genetic mechanism of atrazine resistance in weeds, of which there are now more than 30 species (the resistant weeds began to appear spontaneously in the early 1970s in areas of prolonged atrazine use). Atrazine kills plants by blocking photosynthesis: it is taken into the chloroplast and takes the place of the quinones involved in the transport of electrons. Atrazine resistance is the result of a mutation that prevents the binding of atrazine to the chloroplast membrane where electron transport takes place. The chroloplast genes involved in the mutation leading to atrazine resistance were therefore investigated: the genes in both atrazine-susceptible and atrazine-resistant weeds were isolated, cloned and sequenced. The only difference between the two kinds of genes is one nucleotide base: in the resistant weed, an adenine is replaced by a guanine, and consequently the membrane protein specified by the modified gene does not bind to atrazine (National Research Council, 1984).

In 1982, German and Israeli scientists succeeded in fusing protoplasts of atrazine-resistant *Solanum nigrum* and susceptible potato (*Solanum tuberosum*), but the resulting atrazine-resistant hybrid was unfortunately more like the weed than the potato. Half a dozen laboratories around the world are trying to activate the nucleus in the protoplasts of the donor weed, so that only the cytoplasm of the weed which contains the resistance gene in the chloroplast is introduced into the potato protoplast. From the latter, an atrazine-resistant potato could be regenerated (National Research Council, 1984).

The chloroplastic single gene coding for atrazine resistance could also be transferred into a susceptible plant. This gene has been cloned in a bacterium, and a number of scientific teams are trying to achieve its expression. The greatest obstacle, however, is to discover a vector to carry the gene into a plant and to insert it into the chloroplast (National Research Council, 1984).

The scientists of Calgene Inc. have transferred into a bacterium the gene coding for the resistance to a herbicide (*Round-Up*) marketed by Monsanto. The engineered bacterium became resistant to the herbicide (glyophosphate) and it is planned to transfer the gene from the bacterium into a crop species such as cotton. If this transfer is successful, it would increase the efficiency of the herbicide in cotton fields. The genes for resistance to environmental stresses and to parasites or those controlling productivity are most probably clusters of genes which cannot yet be discovered by the known techniques of genetic recombination. The magnitude of the task of better understanding the genetic constitution of plants can be appreciated when it is realized that approximately 5 million genes are contained in the nucleus of most plant cells. It is not technically difficult to isolate a segment of DNA from a plant cell nucleus, but it is not easy to know which DNA segments are worth using.

Disease resistance, detection and elimination

One of the advantages of *in vitro* tissue culture techniques is the ability to produce virus-free and disease-indexed plants. This is achieved through a combination of thermotherapy or chemotherapy (to remove the virus before or during culture) and tissue culture. Meristem culture makes use of the fact that the meristem remains practically free of virus, even in an infected plant, and meristematic growth outpaces virus proliferation; moreover, vascular connections do not penetrate the meristematic dome of a shoot-tip, which is therefore not easily reached by the virus. A virus-free plant can be regenerated from the smallest explant of a heat-treated meristem placed in culture: a notable example is the potato variety *Belle-de-Fontenay*, which disappeared from cultivation because of viral infection and has been regenerated from a healthy meristem removed from an infected plant and cultivated *in vitro*.

Other techniques include:

— culture of the shoot-apex comprising the apical meristem and one or two leaf primordia

— grafting of the shoot-tip onto seedling rootstocks

— adventitious organogenesis or embryogenesis in special tissues such as nucellar tissue.

The last two techniques are applicable to crops, such as woody plants, for which shoot-tip culture is not possible. As in the case of rapid *in vitro* clonal propagation, virus-free plants have been a commercial reality for over a decade and it is possible, by combining thermotherapy and meristemming, to free nearly all plant materials of nearly all viruses.

Virus-free plants are not, however, virus-resistant plants, because their subsequent exposure to viruses or to virus-vectors when rooted in the soil can lead to their reinfection by viruses, which may occur unduly rapidly because of inappropriate farming practices. Furthermore, plants that are genetically (almost) identical are all equally vulnerable if exposed to disease, with catastrophic results. A virus-free plant production programme must therefore be carried out in conjunction with a rigorous virus-testing programme and with plant protection measures.

In vitro disease elimination methods are particularly valuable because crop yields can be increased significantly, and international exchange of germplasm is facilitated. Such exchanges are needed because crop plant improvement programmes rely on rapidly disappearing wild and semi-domesticated species which supply new genes to produce new cultivated varieties, and virus- or disease-free plant materials provide a rapid and safe means of germplasm exchange across international borders.

In vitro methods of plant propagation also provide a safe means of storing germplasm, either produced *in vitro* or collected in the wild and from less secure field gene-banks (clonal repositories, orchards, plantations, etc.). Conservation methods for vegetatively propagated and recalcitrant seed-producing crop plants include short- to medium-term storage by slow growth for active collections and cryopreservation for base collections.

Recombinant DNA molecules, too, can be used to detect plants with virus resistance. Viruses (including virus-like diseases, or viroids) are insidious pathogens; however, the detailed structure of some of them is understood so that nucleic acid probes can be prepared to be used in screening plant populations for the presence of infective particles. Such a technique has been used for the selection of virus-resistant plants in potato, cassava, groundnut and legumes. To screen new potato varieties for resistance to viruses, breeders traditionally infect the year-old plant in their greenhouse plots with virus, then wait another twelve months to see which varieties show symptoms of infection. This time-consuming process was replaced by immunological techniques, using radio-labelled antibodies and consisting of measuring the reaction with an antigen on the coat of the virus particles in the sap extracted from young tubers; but even immunological techniques are quite expensive.

The gene-probe technology, which is more sensitive and which can process more tests, was developed by scientists in the Agricultural Research Service of the United States Department of Agriculture. They isolated RNA from potato spindle viroid (potato spindle tuber disease for PSTD), synthesized the corresponding copy DNA and cloned this DNA. This riboviroid is hard to eradicate, and its identification in potato tubers made it possible to eliminate the infected organs and to obtain good yields later by using only the healthy tubers. The cloned DNA was an accurate and reliable means of detecting the viroid, as it could be annealed with viral RNA.

Israeli molecular biologists constructed plasmids that contain complementary DNA sequences derived from the *tristeza* virus which is lethal to citrus crop species, causing millions of dollars worth of damage annually to orange and grapefruit orchards. These plasmids, which can be cloned in bacterial cells, have been used to identify minute quantities of the virus in citrus tree barks long before visible signs of the disease appear. By thus uncovering the disease at a very early stage, its spread to other trees can be prevented. Strains of the *tristeza* virus are numerous and not all of them are detrimental to the citrus trees; consequently the Israeli researchers have devised molecular probes to be used in convenient diagnostic kits in order to distinguish among the different strains. Similar test kits for other plants are being developed by commercial firms.

Other serological techniques, which are not sufficiently sensitive to distinguish between pathogenic strain variants, are being replaced by gene probes which hybridize with the pathogen genome and by monoclonal antibodies prepared against the specific antigens. By fusing myeloma cells with antibody-producing cells from an animal immunized against a specific antigen, the resulting hybrid cell ('hybridoma') has the ability to multiply rapidly and indefinitely in culture, and to produce an antibody of predetermined specificity, known as a monoclonal antibody. This hybridoma technology, discovered in 1975, has provided researchers with a tool that permits the production of standardized antibodies (reagents) of a given class, specificity and affinity, as well as the analysis of virtually any antigenic molecule.

For example, certification of pathogen-free potato tubers requires testing for the presence of such organisms as *Erwinia carotovora* var. *atroseptica* (the causative agent of blackleg in potato), potato leaf-roll virus, potato virus X and potato virus Y, which cause devastating losses and reduce yields. The available serological techniques are not sensitive enough to distinguish between strain variants, yet it is essential to determine whether disease outbreaks resulted from variants of the pathogen characteristic of the region where the plant was propagated or characteristic of the region where it was put into production. Monoclonal antibody technology provides a tool for making a distinction between strain variants and for conducting epidemiological surveys of farm land before introducing a given crop plant.

Monoclonal antibodies can replace the available serological reagents made by conventional procedures and that are unable to distinguish between those virus strains causing severe disease and those causing little or no disease. Field diagnosis of barley yellow dwarf virus and barley stripe mosaic virus, for example, would allow detection of strain variants responsible for disease outbreaks as well as determination of whether the disease was seed-borne or the result of field contamination.

In the case of ilarviruses which cause major disease problems in fruit trees (*Prunus* necrotic ring spot and prune dwarf viruses cause disease in cherry trees; apple mosaic and tulare apple mosaic viruses cause disease in apple trees), conventional reagents are difficult to maintain in sufficient supplies to meet the needs. Moreover, the failure of these viruses to elicit a strong immune response has limited the production of reagents to detect the various forms of the virus. Monoclonal antibodies can help overcome such difficulties.

Weed and pest control

The principle behind biocide use is that, in nature, plants and micro-organisms are in constantly evolving competition: researchers are simply attempting to enhance nature's own weapons for use against agricultural pests. For example, bacteria are used to infect and kill insects infesting a field by creating a disease epidemic in an unwanted group of insects. As a bonus, the bacteria expire when they have no more hosts and eliminate the problems experienced with chemical pesticides which tend to linger in the environment. Other plants produce chemicals that discourage plant growth (allelopathic), creating 'living' space around themselves. These genetic traits might be very useful in crop plants.

Perhaps the best known insecticidal microbe is *Bacillus thuringiensis*, a virulent antagonist of many insects. In combination with integrated pest management techniques the use of microbial insecticides can reduce pesticide costs significantly (Orrego, 1981:67-75). Already UNDP/FAO is funding projects to use *B. thuringiensis* to control olive pests in Greece and alfalfa pests in Argentina (Orrego, 1981:68). A number of developing countries have evolved *B. thuringiensis* preparations for use against their particular insect problems. Recently the gene coding for the toxin of *B.thuringiensis* has been successfully transferred into tobacco plants, though it remains to be seen whether the plant cells produce the toxin in sufficient quantities to kill the pests (Yanchinski, 1985).

There can be no doubt that the entire field of biocides is only in its infancy and will surely grow during the next decade. In the case of useful bacteria and viruses the research effort need not be highly sophisticated and can be conducted by teams of entomologists and microbiologists, and effective methods can be developed at relatively low cost. The entire field of bioinsecticides (and integrated pest management) offers great potential for lowering the costs of pesticide imports significantly while also decreasing environmental damage.

Biological nitrogen fixation

The possibility of biological nitrogen fixation has received much attention in the last ten years because of the enormous costs farmers are incurring in the purchase of nitrogen fertilizer. The problem is not that there is insufficient nitrogen (80% of the atmosphere is nitrogen), but rather that this gaseous form cannot be assimilated by living organisms. There are, however, strains of bacteria (such as *Rhizobium*) and algae that can convert atmospheric nitrogen into a form usable by plants, a process termed nitrogen fixation. In legumes, the bacteria live in root nodules and are provided with nutrients by the plant's vascular system and in return provide fixed nitrogen to the plant. The value of nitrogen fixed by legume symbionts has been estimated as worth nearly $10 billion in American agriculture alone (Orrego, 1981:51).

Research is being undertaken to discover ways of improving and extending biological nitrogen fixation by manipulation of the appropriate genes (*nif* genes) (Postgate, 1982). If the genes could be transferred from *Rhizobium* to other bacteria, it might be a means to extend nitrogen fixation to crop species other than legumes, in particular to grains. The technical difficulties of genetically engineering symbiotic nitrogen fixation are enormous: for one thing, there are at least seventeen genes involved in symbiotic nitrogen fixation which would have to be transferred in a cluster. Then, the final expression of the genes must be carried out in the plant cell, but the mechanisms involved are not identical in the bacterium and in the plant, so that two interacting organisms must be engineered. Nitrogen fixing genes have been successfully transferred from *Rhizobium* to *Agrobacterium tumefaciens*, which then infected an alfalfa plant but the genes were not expressed (National Research Council, 1984).

Quite apart from the technical problems, questions have been raised as to whether yield losses arising from nitrogen fixation energy demands would be economically prohibitive. A corollary question is whether photosynthesis in plants can be made efficient enough to compensate for the energy used in nitrogen fixation. In fact, engineering corn to fix nitrogen could lead to yields at least 10% lower than those currently achieved. Plants and bacteria engineered to produce nitrogen symbiotically at an economic price is probably far in the future, and even genetically 'improved' free-living soil nitrogen fixers probably may not be successful in competing with the naturally occurring soil bacteria or fertilizers derived from natural gas.

Another area of very active research is to improve the capability of free-living nitrogen-fixing bacteria that secure energy from decomposing matter and fix nitrogen into the soil. Research is directed to making these more efficient and encouraging their growth in the plant's rhizosphere so that it can utilize the newly fixed nitrogen.

The most promising research area for reducing fertilizer consumption in the next ten years is the further development of farming systems using blue-green algae and *Azolla* (Jagannathan et al., 1978). The *Azolla* ferns provide energy and living sites for the algae, and the algae produce nitrates which the *Azolla* absorbs. Rice paddies fertilized with composted nitrogen-rich *Azolla* report yield increases of 12-14% over control plots. Usage of *Azolla* is widespread in Asia, and in California it is in some cases providing up to 75% of the paddy fields' nitrogen requirement (Orrego, 1981:58). Nevertheless, although research is forging ahead and a number of research institutions are committing research funds to biological nitrogen fixation, success in fields other than *Azolla* usage is probably relatively remote.

The reason for great emphasis on biological nitrogen fixation research in developing countries is obvious — foreign exchange savings. Whether, with the exception of the *Azolla* research, biological nitrogen fixation is a well-placed investment is certainly open to question. The problems include: (1) normal soil conditions may not provide free-living nitrogen fixers with the required energy sources; (2) the nitrogen that is produced will not be directly absorbed by the plant, thereby rendering the process relatively inefficient; and (3) ammonia may be overproduced, rendering the soil acidic. At first sight, biological nitrogen fixation appears to be an ideal strategy since fertilizer can be produced by farmers at very low cost, but if the research does not come to successful fruition, developing countries will have squandered much of their research resources on a gamble that has already been abandoned after major investments by multinationals.

Other agricultural and food applications of biotechnologies

ANIMAL HUSBANDRY

The 'biorevolution' will not be confined to plants and, in fact, will probably initially have as great or greater impact on animal husbandry. There are three distinct areas in which biotechnology will affect animal production: the use of the new reproductive technologies, new vaccines and bacterially produced hormones. These techniques are in principle applicable to any animal, but the bulk of the research is directed towards cows, swine and chickens.

New techniques to control animal reproduction are already on the market in developed and some developing countries. The animal having the largest market potential and receiving the greatest research attention is the cow. Bovine artificial insemination has been available for many years, but in the last ten years scientists have developed techniques for non-surgically transferring embryos from one cow to another. This is coupled with the use of hormonal preparations to induce the cow to superovulate, i.e. to release simultaneously up to twenty ova from its ovaries, whereupon the cow is artificially inseminated. After six days the fertilized embryos are removed and transplanted into 'surrogate' mothers using non-surgical embryo transfer techniques which are already routinized and performed by farmers in developing countries. These techniques make it possible to increase greatly the number of elite dairy cows that can be reproduced.

Embryo transfer is leading to other services. Until recently, the embryos had to be transplanted into the surrogate mother within one or two days, but freezing techniques have improved sufficiently to provide a 30% survival rate and embryo transfer is becoming commercially viable. An American company, Genetic Engineering Inc., has developed a technique for sexing embryos before implantation, which allows the dairy farmer to select females and the cattlemen to select males, saving the cost of bringing an unwanted embryo to term. It is also possible to split or twin fertilized ova (at this point up to sixteen identical clones can be produced). Thanks to the freezing techniques it is possible to raise an adult cow, measure its milk production or weight gain and, if production is satisfactory, the elite cow's identical twins can be thawed out and implanted in a surrogate mother.

For developing countries, these new animal reproduction techniques offer increased flexibility and opportunities. For example, since embryos are not subject to quarantine and of course are not as large, unwieldy or expensive to transport as one full-grown cow, it is possible to move hundreds of embryos from elite parents in a suitcase-sized freezer to any location. As an added bonus, the surrogate mother provides environmental immunities to the young calf, thus drastically reducing mortality rates for cattle imports. The speed with which a cattle herd could be upgraded is greatly increased, and purchases of elite germplasm are inexpensive in comparison with the cost of adult cows.

These techniques are not limited to cows. Researchers in Kenya have successfully bred from embryos of the West African N'Dama, brought frozen from the Gambia. N'Dama are immune to trypanosomiasis and so could make a valuable contribution to pastoralism in tse-tse fly infected areas. Indian scientists have had similar successes with embryo transfer techniques applied to buffalo and goats.

Another major obstacle to animal husbandry in the developing world has been the number and seriousness of animal diseases. Biotechnology provides numerous opportunities to attack animal diseases, and vaccines are currently being developed against bovine and porcine scours, pseudorabies, bovine adenovirus and viral diarrhoea, foot and mouth disease and Rift Valley fever, among others. Vaccines could obviously provide important economic benefits, but as is so often the case, the animal diseases first tackled will be those that are the most profitable — not those endemic in the developing world. Animal vaccines are so early on the market for two reasons: firstly, that biotechnology expertise was developed in medical schools and biology departments engaged in research on human vaccines, much of the knowledge of which could be easily adapted to animals; secondly, that animal health products, though similar to human health products technically, require a far shorter testing period and receive rapid market approvals with the result that many companies have emphasized animal health products.

For developing countries, the single most commercially important vaccine is that being developed against foot and mouth disease (FMD), a virulent and economically devastating viral disease that infects cloven-hooved animals. The costs of FMD eradication efforts are extremely high — the elimination of an FMD outbreak in England in 1967-8 cost in excess of US $200 million and a similar Canadian outbreak in 1952 was estimated to have cost US $1 billion (Blackwell, 1980:1019) — so that uninfected countries are obliged to embargo meat shipments from infected regions such as Argentina, Brazil and most of Africa to the United States of America and Japan. The market for a safe, effective and easy-to-handle FMD vaccine is vast. Argentina, for example, imports 200 million doses per year of the attenuated virus vaccine from the United Kingdom. The traditional vaccine must be administered every three months and, in certain cases, has actually caused FMD outbreaks (Blackwell, 1980), whereas the genetically engineered vaccine cannot cause disease outbreaks and should be more effective. Several companies are pursuing a genetically engineered FMD vaccine, though none has yet developed an entirely effective one.

The impact of a totally effective FMD vaccine could be a substantial increase in meat exports by developing countries to the industrialized nations because the barriers to meat entry into the FMD-free countries could be dropped. The increased exports could easily increase the value of grazing land and perhaps cause a shift from food cultivation to animal grazing. Each of the vaccines listed above offers the potential of increasing efficiency and productivity in animal husbandry, and just as surely offers the possibility of shifting relative factor prices against poor farmers and consumers in developing countries, perhaps increasing relative or even absolute poverty.

Genetic engineering has made it possible to produce bovine, porcine and chicken growth hormones and bovine interferon. Bovine growth hormone has been shown to increase a cow's milk productivity by between 10 and 20% (Peel et al., 1981), and similar results have been achieved with chicken growth hormone in speeding the growth of broilers (Boone et al., 1983). The actual utilization of these growth hormones is blocked, however, by the lack of an adequate delivery system — the hormones are metabolized in the digestive tract when taken orally. Bovine interferon is being tested for prevention of shipping fever, a disease that results in severe weight loss in up to 30% of cattle housed in feed lots. The increasing number of animal agriculture inputs will make animal production more efficient for those able to afford these inputs.

SINGLE-CELL PROTEINS AND MICROBIAL CONVERSIONS

The engineering of a micro-organism's metabolic pathways makes it possible to convert efficiently low-value feedstocks into higher-value products such as amino acids, proteins and specialty chemicals; in certain cases the products of these altered micro-organisms will compete effectively with agricultural production. Much research money is being invested in developing micro-organisms capable of transforming agricultural wastes consisting largely of cellulose and lignin into higher value products, though researchers have yet to make the process economic. Some of the applications currently being investigated include high fructose corn syrup, papain and conversion of palm to cocoa oil; there will undoubtedly be many more as increasing numbers of food processing giants enter the industry.

The term 'single-cell proteins' designates microbial proteins produced by the mass culture of yeasts or bacteria on hydrocarbons or other substrates, often waste by-products with few alternative uses. After a setback following the large increase in oil prices in the 1970s, the microbial production of SCP has received reviewed attention from large petroleum and petrochemical companies because of protein's relatively high market value. Imperial Chemical Industries, for example, has produced an SCP animal feed but the facility has yet to become competitive with soybean protein. The ICI process feeds yeast on methanol and requires approximately the same amount of fossil fuel energy as soybean production, while using only one-tenth the labour; capital investment is, however, very high (Yanchinski, 1981). If increasingly efficient yeast strains and process engineering are developed, the SCP process may become cost-effective. For oil-exporting countries, the production of SCP may make economic sense even earlier because it can be derived from 'free' natural gas — the successful entry of OPEC countries into the world protein markets would certainly disrupt the agricultural economies of countries that export cattle protein feeds (soybean cake: United States of America, Argentina and Brazil; peanut cake: Senegal and other West African countries). However, SCP has not yet had even a modest economic success.

Genetic enineering offers the possibility of increasing the yield of cyanobacteria such as *Spirulina* which have been exploited for centuries in Mexico and Chad as a protein source. Many countries have experimented with proteins derived from *Spirulina* for use in animal feed and as a dietary supplement for human consumption.

An area of substantial growth is the fermentation of the amino acid lysine as a cattle feed supplement. The two major world producers, Ajinomoto and Kyowa Hakko Kogyo, have recently constructed new amino acid production facilities in the United States of America, and both are actively using genetic engineering to develop more efficient micro-organisms. Lysine is an example of the contradictory impacts biotechnology could have: genetic engineering of micro-organisms could make lysine production less expensive, but other research in genetic engineering could develop corn plants that produce sufficient lysine to eliminate the need to add lysine to cattle feed. Research, investment and success in either option will affect the structure of agriculture differently.

French researchers are undertaking to develop microbial techniques to transform inexpensive oils such as corn oil into more valuable oils such as that of cocoa (Cantley and Sargeant, 1981:331). In this case many plant oil exporters in developing countries could be seriously affected by competition from companies based in industrialized countries in the international food oils market.

Examples of other techniques under development include one Japanese company's plan to develop bioreactors on board vegetable oil tankers. These would process the food oil even while *en route* to Japan, decreasing the necessity of investments in food oil processing plants in developing countries and shortening product turnover time.

FERMENTATIONS

Many developing countries have traditionally used fermentation processes in order to increase the nutritive value of their agricultural products and preserve them. Biotechnologies allow these techniques to be improved and extended, making the established process more efficient and also creating new products.

Research conducted jointly by ORSTOM and IRCHA in France, and subsequently pursued in Mexico, has led to the protein enrichment of cassava flour by growing in it filamentous fungi. An *Aspergillus niger* strain is grown on the flour (best results have been achieved with 55% water content and a temperature between 35°C and 40°C), and after 24-30 hours the fungus increases the protein level of the cassava flour from 2 or 3% to 18-20% and produces an acceptable and enriched animal feed. Similar techniques of protein enrichment have been developed for other starchy tubers, using a variety of fungi. The process is easily adaptable to small-scale installations at the village level.

Related methods can be applied to the improvement of foods for human consumption. Furthermore, the enzymes synthesized during the growth of *Aspergillus niger* could be isolated and used in various processes in the food and chemical industries. The potential applications of such microbially produced enzymes are extensive.

In Cuba, since 1965, ten plants have been producing approx. 100,000 tons of fodder-yeast per year (1985-6) by growing the yeast *Candida utilis* on molasses supplied by the sugar industry; the yeast biomass is used as animal feed. Cuban researchers intend to improve the economics of the process through using other by-products and residues of sugar-cane cultivation, and through treating the yeast so as to lower its nucleic acid content, hence making products suitable for human consumption. Finally, mixtures of wet yeast and molasses can be used also as animal feed, and thus avoid the energy-expensive process of drying the yeast.

In general the trend is towards the blurring of the distinctions between the physical, chemical and biological elements in raw materials, so that they become increasingly interchangeable. Reliance on traditional crops to produce specific products is likely to diminish, with adverse effects for food exporters in the developing countries as the industrialized countries reduce their dependence on natural raw materials. On the other hand, the new technologies may, in the long run, permit developing countries to improve the nutritive value of their food supplies and derive food from completely unsuspected sources.

Agricultural biotechnologies in the developed countries

The industrialization of agricultural biotechnology in the developed countries has happened relatively recently, rather later than similar developments in the biomedical fields. As in biotechnology generally (and as also in the field of information technologies), there has been a close linkage between research institutions (particularly universities) and commercial firms that can develop and market the findings of researchers. The most common method in the United States of America has been for the researchers themselves to found specialized agricultural biotechnology firms in partnership with entrepreneurs; some of the existing general biotechnology firms have also branched out into agriculture-related R&D programmes. In addition, multinational corporations in pharmaceuticals chemicals with interests in agricultural chemicals and seeds have been eager to buy expertise, either from the small specialized companies or directly from the universities.

Capital is a critical requirement in such enterprises because the major products of biotechnology are unlikely to be available immediately: there may be a delay of as much as 5-10 years before a marketable product is produced, falling to as little as 2-3 years for companies using monoclonal antibodies. In 1980, the start-up costs over the first three years for a (non-agricultural) biotechnology firm was estimated to be roughly US $10-12 million and a staff of 25 PhDs for the R&D stage alone, though some agricultural biotechnology firms have been launched for less. As in the field of computers, venture-capital funds have been an important source of finance, particularly at the very outset. Once established, firms may seek further capital from stock offerings, and also from research contracts with the multinationals — which puts the research company in the awkward position of selling the knowledge crucial for its own survival to its competitors.

The involvement of the universities in these commercial ventures has raised issues of conflict of interest and about the openness of research findings that had not previously troubled academic research. The university is a very special institution in any society because it trains the skilled labour force of the society and does much of the basic research from which new productive forces are derived. The information derived from this research has traditionally been communicated openly to all interested in acquiring the knowledge, an openness that has provided an access to science unavailable from private companies. The free flow of information and lack of pecuniary motive has allowed scientists and students from any country to come and learn in a relatively open and free environment, and has provided a medium for technology transfer to developing countries.

The new links between universities and commercial interests have altered this situation. Will university-based researchers continue to pursue knowledge in a reasonably disinterested way, or will they skew their research to satisfy their commercial patrons, for example by developing plant varieties resistant to specific proprietary herbicides? Privatization has been encouraged by university administrations, who accept corporate funding for laboratories and research teams (e.g. University of Illinois' US $2 million plant biology centre funded by Sohio, or the work of the Australian National University in research on nitrogen-fixing soybean varieties financed by Agrigenetics). European academics have been less entrepreneurial than their American counterparts in setting up companies themselves but are faced with similar pressures from commercial interests.

The new biotechnologies emerged at a very propitious moment for the large chemical pharmaceutical companies, many of whom have been experiencing a pronounced dip in profits and a slowing of the rate of new drug and chemical discoveries (Kenney, forthcoming:300; Steward and Wibberley, 1980). Biotechnology provided the opportunity to abandon commodity chemicals and move into more lucrative specialty and agricultural chemicals. In pursuance of this strategy, and to strengthen their position in agrichemicals, the multinationals have been purchasing seed companies. Since the seed contains the DNA programme that a living plant follows (i.e. the seed determines all of the valuable traits of the grown plant, such as stress tolerance, plant morphology, yield potential and responses to agrichemical inputs), seed research may yield chemical-seed packages that would oblige farmers purchasing seeds to use certain proprietary chemicals, thus ensuring a market for the company's products. For example, Ciba-Geigy markets a sorghum seed-chemical package in the Sudan: the seed is coated with a patented 'safener' to protect it from Ciba-Geigy's own patented herbicides. New biotechnologies will also improve techniques for producing hybrids, which have reproductively unstable progeny and so cannot be replanted by the farmer, who must therefore return to buy seed annually, thereby creating a captive market. Furthermore, successful hybridization of wheat and rice may open up enormous new markets for seed subsidiaries of multinationals.

Reference has already been made to the role of the multinationals in funding research in universities and specialized firms (for some American examples see Table 6); they have also invested in large in-house research facilities in their efforts to master the techniques which potentially have such important consequences for markets in their existing products as well as in completely new seeds and chemicals.

In Europe and Japan, government initiatives and support have played a crucial role in the development of biotechnology. In addition to the European national programmes, the European Economic Community invested 4 million ECU per annum from 1982 through 1986 on various projects in Europe. In 1984, the EEC adopted a new five-year plan for biotechnology that provides for: research and training, concerted action on monitoring and communicating results, securing access to raw materials, standardizing regulatory regimes inside the Common Market, standardizing intellectual property law, developing demonstration projects. The entire budget for the EEC effort is expected to be 35 million ECU for the first two years. These totals are in addition to the major sums each country is allocating to biotechnology research. Long-established seed, fertilizer and horticultural companies and also recently created specialist firms are applying the latest techniques to the development of a wide range of new cultivated varieties, micropropagation of plantlets and gene transfer.

The Soviet Union is also very active in research, and has achieved successes in a variety of areas (Butenko, 1984). Much work has been done on the potato: somaclonal variants and somatic hybrids have been used to

Table 6. Some agricultural biotechnology start-ups: selected university-based researchers, financial linkages, and areas of research.

Company	University affiliation	Financial linkages	Areas of research
Agrigenetics	University of Illinois University of Wisconsin Cornell University	purchased by Lubrizol	seed-related biotechnologies
Advanced Genetics Science	Harvard University University of California, Berkeley	Rohm & Haas Hilleshog	cloning of disease-resistant potatoes
Calgene	University of California, Davis	FMC Corporation Continental Grain	plant genetics
Genetic Engineering Company	University of Colorado Ohio State University	Johnson & Johnson	animal reproduction
DNA Plant Technology Company	Texas A & M Columbia University University of California, Berkeley	Campbell Soup (40%) Koppers Co. General Foods	tomato, tobacco, forestry products
Molecular Genetics	University of Minnesota NIH	American Cyanamid Moorman Manufacturing U.S. Dept. of Defense	corn, scours, prevention and nonagricultural applications
Sungene	Washington University CalTech	Lubrizol Mitsubishi Aambrecht & Quist	sunflower corn hybrids

* All information is accurate to the best of our knowledge, but it should be kept in mind that the proprietary nature of these firms makes it difficult to keep abreast of the latest data.

Source: Buttel et al., 1983.

increase yields, resistance to frost, to virus diseases and other pathogens. Widely applicable techniques have been developed for the selection and establishment of cell-lines, the indexing of virus-free plants, and the speeding up of other plant-breeding processes. Cryopreservation is used to assist *in vitro* cultivation and to store banks of successful cultures.

The Japanese experience in biotechnology may provide a useful model for the less-developed countries. Early developments were based on traditional skills and interest in fermentation processes. There has been a history of steady expansion, pooling the resources of government, universities and major companies; in addition, the Japanese have sent researchers for training in the United States of America and Europe, and have not hesitated to buy technologies from foreign firms. As a result, Japan is now one of the leading countries in the field of biotechnology research and production.

Some of the developments in industrialized countries could clearly have highly undesirable consequences for the developing world. Participation in research may become more difficult if the companies who fund research in universities protect the findings with trade secrecy and patents. Knowledge, insofar as it is made available, is likely to be expensive. As consumers, farmers in developing countries are going to be increasingly vulnerable to the research and marketing strategies of the multinationals. These trends could be even more difficult to resist if agricultural research activities in the United States of America are centralized, as advocated by a joint White House Rockefeller Foundation report, *Science for Agriculture* (1983), a move strongly supported by the multinationals, since it would make it easier for them to control the agenda for research. Other developments are more encouraging — as, for instance, the establishment by the United States Department of Agriculture and the University of California of a Plant Gene Expression Center which will make its results available to all and will refuse industry support that involves secrecy and exclusive rights.

Impact of biotechnologies on developing countries

BENEFITS FOR AGRICULTURAL PRODUCTION

Many of the advantages of the applications of biotechnologies have been enumerated in the course of the technical discussion, and potentially they apply equally to developing and developed countries. Plant breeding can be carried out more rapidly and accurately, creating high-yield varieties that can be adapted to grow even in adverse conditions, while similar improvements obtain in animal breeding. Tissue culture can be especially useful in propagating trees that require long maturity periods, for example. A wide range of characteristics can already be selected for and then deliberately developed; as techniques are refined, this range could be extended almost undefinitely. The most valuable aspect of biotechnologies for food production is that they provide techniques for increasing yields, especially of the coarse grains and legumes that make up the bulk of the world's food. Traditional fermentation processes can be improved and extended. The new technologies also permit transfer and storage of materials for plant and animal breeding without the normal problems of disease and contamination. Disease and pests can now be attacked by non-chemical means, with beneficial effects on the environment.

Plant tissue culture can bring improvements to production of many useful substances derived from agriculture, such as pharmaceuticals, dyes, flavourings and fragrances. Production by these new techniques can be continuous and not limited by season or climate. The quality is more reliable and can be controlled more easily — moreover, 'natural' products are viewed as safer than manufactured chemicals. Installations do not need to be designed to cope with harvest peaks of production and storage: they can function year around and be located close to consumers and markets instead of being tied to cultivation sites. Conventional agricultural production — which inevitably exhibits varying quality and yield owing to environmental conditions and farming practices — can be complemented with plant biotechnology products.

While some of the research and the resulting techniques involve expensive equipment and highly trained personnel, many of the biotechnology products can be developed and produced in facilities that are not very sophisticated, and many techniques can be adapted for use in developing countries.

ADVERSE FACTORS

Costs

The initial costs of the new techniques and products are, however, usually rather higher than traditional ones because of the long and expensive process of research and development. Inevitably this means that many of the benefits of biotechnologies will be beyond the reach of the farmers in greatest need, while the large growers will be able to afford the new products and hence will place the small producer at an even greater disadvantage. The capital outlay required for research facilities is now so large that the only feasible way for developing countries to participate is by negotiating cooperative agreements with interests in the industrialized countries, and they are likely to be forced to pay dearly for knowledge and products now that the stakes have increased so dramatically.

Displacement of traditional products

Plant tissue culture offers increased possibilities of substituting industrially produced inputs for agricultural commodities, and many high-value plant-derived products are vulnerable to displacement as a result of current research. At present the production costs are sufficiently high that plant tissue culture methods provide a viable alternative only for products valued above US $600 per kg (Tudge, 1984), but increasing research and production experience will allow these costs to be reduced significantly. The impact of successful production of substitutes could be catastrophic for countries dependent upon exports of the natural products, in which often they have hitherto enjoyed a virtual monopoly because of the specific conditions required for growth. Displacement has been happening for a long time as synthetics have been developed (e.g. the Indian indigo industry, jute in Bangladesh and steroids in Mexico); plant tissue culture is a far greater threat because the techniques can be applied to any plant, whereas there are technical difficulties in synthesizing many plant products by conventional synthetic techniques. Active research is currently underway to produce by cell culture the products extracted from opium, chinchona, digitalis, ginseng, pyrethrum, tobacco, lithospermum and the active ingredients of chilli and saffron (Buttel et al., 1983).

Producers of the displaced products must prepare for the dissolution of their monopolies, either investing profits — while they last — in other industries or rationalizing production so as to be more competitive. Crop improvement techniques could provide the means to do so: an example of the double-edged impact of biotechnologies.

Employment and social effects

As in the case of other technologies, the impact of the new agricultural technologies will displace workers in some areas while creating opportunities in others. If the new seeds result in the cultivation of previously marginal lands, now viable in spite of high aluminium content or waterlogging, etc., the existing agriculturalists will be enclosed as land values rise, as happened in the previous Green Revolution (Pearse, 1980). The new technologies allow output of traditional crops to be increased without any increase in employment, and new varieties may require less attention and therefore a smaller workforce, though these negative effects may be offset (perhaps elsewhere in the economy) by the higher returns — if they are spent in the producer country. As we have seen, too, plant tissue culture techniques create capital-intensive substitutes for other products with concomitant decrease in employment in the affected sectors. The benefits of plant biotechnology, such as higher-yielding clones or *in vitro* vegetative propagation of superior cultivars, will in all likelihood strengthen the strongest economic units and the large estates, rather than the small farmers. Such a situation will result from the fact that the large estates have the operational skills, financial resources and market experience to take full advantage of the new techniques and know-how. Smallholders could even be displaced as a consequence of such a trend, unless appropriate national policies organize the equitable distribution of the benefits to the small plantations as well.

Privatization of knowledge

Allied to the enormous costs — and lucrative markets — involved in the new techniques, is the growing desire to protect acquired knowledge and its products which makes access to essential information both difficult and expensive for those on the outside. Patent law is not yet entirely clear on the right to patent micro-organisms created by genetic engineering or more traditional bacteriology, which leaves the situation somewhat more open for the time being. Patents in any case at least make the knowledge available at a price (which may even be waived amongst university researchers, as in the case of Stanford University policy on royalties: see Sasson, 1984); commercial secrecy can be even more problematical. A particularly ironic twist for developing countries is the case of patented varieties derived originally from their own phytogenic resources for which they must now pay. Some developing countries are trying to protect themselves against future exploitation of this kind by banning exports of germplasm.

Monopoly power of suppliers

The enormously profitable markets involved, as well as the high R&D costs, have inevitably attracted the interest of powerful multinational companies who are currently in a dominant position as regards biotechnology products. Their control over research and marketing means that the applications of agricultural biotechnology techniques may tend to favour the improvement of crop species and the upgrading of plant products almost exclusively to meet the needs of the international market. For instance, the dessert banana will be favoured over bananas used locally for cooking (plantain), coco-tree improvement will be oriented towards selecting clones that produce high yield of coco-oil for the margarine, soaps and lubricants intended for the overseas market rather than towards the multipurpose uses of coconut in the local economy. The development of specific herbicide-resistant varieties and of hybrids generally (which require annual purchases of seeds) will give major seed companies increasing control over consumers.

Erosion of genetic resources

The last ten years have led to an increased awareness among agricultural scientists of the erosion of the world's genetic resources, i.e. genetic variability that different plant varieties have encoded in their DNA. For instance, the many varieties of corn have different qualities — yield, stature, pest resistance, moisture tolerance, etc. — and those differences reflect differing capabilities of response to environmental variation which plant breeders can use as raw material for the breeding process. Obviously, the larger the pool of varieties available, the greater the inventory of traits available. When a greater number of varieties with different characteristics are planted, not only do you have more information, but also the total crop planted has greater resilience. As an example, if there is a drought, at least the drought-resistant varieties survive. The same is true regarding pest infestations. Conversely, the monoculture of a single variety creates a more fragile system, i.e. if this variety succumbs to an environmental tort, each of the genetically identical plants is affected, posing the possibility of an epidemic. The tendency towards standardization of varieties with desirable traits may be somewhat offset by somaclonal variations generated by plant breeding activities.

The major centres of plant genetic diversity are located in the developing world, which is therefore the source of genetic materials for the food we produce, export and consume. The modernization of agriculture is creating a situation in which single genetically uniform varieties are being sown over large acreages and displacing traditional varieties. The response to the problem of genetic erosion has been the collection and storage of traditional seed materials in germplasm repositories. In recent years, developing countries, feeling that their resources were being unfairly exploited by the developed nations, have begun to demand some form of compensation and to forbid germplasm collection in their territories.

Response to 'biorevolution'

The biorevolution offers many opportunities to developing countries to secure real economic gains, but securing these gains will require more than simply lavishing money and equipment on scientists and hoping for success. Fundamental to long-term success is a programme of elementary and secondary education dedicated to giving students first-hand scientific experience. Plans must be based on an assessment of national resources and needs, and must aim towards realistic goals. Applied research must be directed in such a way as to maximize its relevance to the needs of the country, reinforced by analysis of what is technically and economically feasible.

There is no single correct strategy. For example, India with its large research budget and pool of trained manpower will necessarily have a different approach from that of, say, Zimbabwe. Yet, even the smallest and poorest countries can participate in the biorevolution through regional and international networks. Participation need not be in the most sophisticated basic research, but rather should match national skills and research expertise with problems that can be tackled. The greatest mistake would be to attempt to mimic the research agendas of the developed countries.

The two major difficulties are costs and personnel, since many biotechnology developments require both capital and highly trained scientists. On the other hand, the costs are not high in comparision with the costs of imported weapons or luxury car assembly plants, for example, nor is it necessary to aim for the high-cost end of the spectrum. As regards personnel, too, the bottleneck is often in provision of appropriate local opportunities for trained scientists in order to stop the significant exodus of qualified people, who already exist but emigrate because of lack of funds and facilities for research. Another important but neglected aspect of manpower needs is the availability of trained technicians capable of maintaining and repairing scientific equipment.

Developing countries also face problems of infrastructure: the need for dependable electricity and water supplies, and efficient transportation. A prolonged electric power cut could result in the loss of months of work. Certain enzymes and reagents are very unstable and must be delivered in a frozen state within 48 hours if they are not to deteriorate (Baltimore, 1982). Delays in repairing and servicing imported machinery can be disastrous. The entire problem of inadequate infrastructural development may be even further reinforced during periods of economic or political uncertainty when the immigration and customs service can break down almost completely.

Another critical factor is political will and resolve, which is essential in order to overcome some of the difficulties enumerated and to sustain policies once adopted. An investment in biotechnology is not a one-time investment, but rather requires constant and growing investments as projects come to fruition. The prospects for immediate riches such as are realized in industrialized countries when the stock is sold to the public are not possible in developing nations. False expectations of the ease of success, sometimes fostered by the scientists themselves, will rebound to the detriment of creating a viable biotechnology industry.

SOME NATIONAL PROGRAMMES

A number of countries have launched national biotechnology programmes, including Cuba, the People's Republic of China, Brazil, Argentina, Thailand,

India, Algeria, the Republic of Korea and the Philippines, while many more are making investments in this area. The undertakings of two very different countries, India and Cuba, are examined in detail, outlining the strengths and weaknesses of their respective efforts; the programmes of other countries are also briefly discussed.

India

India will undoubtedly have some success in its biotechnology effort thanks to its large industrial base and extensive cadre of local and émigré educated scientists. An important (and unpredictable) variable affecting the success of the Indian effort will be the degree of commitment that the important private sector assigns to biotechnology. (There are encouraging signs from the joint venture agreements between Tata Co. and Japanese and American firms, which will use biotechnology for tropical crop improvement).

A National Biotechnology Board (NBB) has been set up to organize the national effort, and has developed its 'Long Term Plan in Biotechnology for India' (1983) which formulates specific proposals meant to ensure India's participation in biotechnology; the NBB will encourage and facilitate interdisciplinary research and build the infrastructure necessary to support research. For example, the government has undertaken to import and supply necessary enzymes and biochemicals and is setting up production facilities for the most important of these (NBB, 1983:30). The government is also giving high priority to the import of needed equipment. In the personnel area, the plan calls for the training of 50-100 'biotechnologists' annually. On the other hand, no mention is made of training technicians to repair equipment and assist in other activities crucial to laboratory success.

India is rich enough in universities and research institutes that it can do any but the most state-of-the-art research. The large number of researchers, stable foreign exchange situation and excellent universities permit infrastructure building which most countries could not afford or support. The government has shown its commitment to develop a large-scale programme that addresses all aspects of creating a biotechnology industry; the success of the programme will contribute significantly to Indian agricultural and industrial productivity. Another major contribution to the Indian biotechnology effort was the decision by UNIDO to locate one centre of the International Center for Genetic Engineering and Biotechnology (ICGEB) in India.

Cuba

Cuba, despite a much smaller population and industrial research base than India, is making a determined effort to participate in the 'biorevolution'.

R&D goals are strongly influenced by the specific development needs of the economy, including the commitment by the state to improve medical care, the importance of agriculture (sugar and tobacco) to the economy, the political commitment to raising animal protein consumption in the diet, the desire to be an exporter of biomedical products, and strong research capabilities in the agricultural and medical sciences. Cuba does not, however, aim to be a scientific leader in world biotechnology — rather, the Cubans wish to take advantage of the results of research conducted overseas by remaining somewhat behind the cutting edge of advanced science, yet in a position to exploit commercial aspects of scientific developments.

Cuba launched a biomedical genetic engineering laboratory with, as its first project, the production of interferon using the Cantel process; laboratories are also examining the possibility of using recombinant DNA to produce alpha interferon. In pursuit of the project, Cuba has held two international interferon and biotechnology seminars and launched a journal entitled *Interferon e Biotecnologia*. There are several reasons for tackling interferon: to provide a model system for learning recombinant DNA techniques and to derive immediate medicinal benefits (Ubell, 1983); furthermore, the project is an impressive demonstration of national technical expertise. Other major projects underway are the use of tissue culture to improve sugar cane, and the production of animal vaccine. There has also been success in bovine embryo transfer, and Cuban scientists claim to have successfully extracted lysine from yeast and already introduced the method into local industry (Ubell, 1983).

Cuba is launching an expensive product-focused drive to introduce biotechnology into the economy, but deliberately without trying to 'ape world class research' (Holtzman in Ubell, 1982). They have chosen targets that can benefit the ordinary individual. The success Cuba has had in building a significant advanced biotechnology research establishment clearly demonstrates that political commitment and desire can overcome many obstacles (Beardsley, 1986).

Some other national programmes

Another example of a major biotechnology effort is the Mexican national programme, the centrepiece of which is the US $3 million Research Center for Genetic Engineering and Biotechnology located at the University of Morelos in Cuernavaca, and dedicated to doing very sophisticated genetic engineering. In addition, a number of other researchers are involved in much more applied plant tissue culture research for agriculture (Robert and Loyola, 1984). The variety of projects being funded in fact poses some very serious questions regarding the allocation of resources between expensive basic research and the less expensive applied research such as plant tissue culture; each society must decide where its priorities lie (Quintero, 1984).

So, for example, in Ivory Coast, a biotechnology laboratory has been set up by the Institut français de recherche scientifique pour le développement en coopération (ORSTOM). The central unit is equipped for *in vitro* plant propagation, while three satellite units are focusing on (1) genetic improvement of plants (somatic hybridization) and conservation of medicinal plant germplasm; (2) phytopathology and protection of plants against virus and nematode diseases; and (3) production of high-value secondary metabolites (Marin, 1984). The laboratory is dedicated to the national exploitation of the plant genetic heritage of the Ivory Coast and to designing a commercial policy for marketing plant biotechnology products. It represents a significant technological advance for Africa.

There are even simpler efforts than those discussed thus far that can be undertaken, such as clonal propagation of certain agriculturally important species. In the Socialist Republic of Viet Nam, for example, family-size potato tissue culture facilities have been established which reproduce and distribute to farmers elite potato varieties such as those developed by the International Potato Center (CIP) in Peru. A single facility can produce 100,000 clones per month, rapidly speeding the diffusion of new varieties. The Vietnamese example illustrates that biotechnology can be adapted to meet many different goals. Other countries as varied as Nicaragua and the Philippines are developing low-cost techniques for microbial pesticide production. In many ways biotechnology success in the developing countries is limited only by the dedication and imagination of their scientists.

INTERNATIONAL COOPERATION

The traditional routes for developing countries to gain access to agricultural technology and its products were, on the one hand, to import or produce under licence from multinationals items like machinery, fertilizers and pesticides, and on the other, to obtain information on plant breeding, animal and soil science from universities and research institutes, at minimal cost or even *gratis*. The strong links developing between business and research interests threaten to disrupt this access to knowledge. Large firms rarely license anything but older and less profitable technologies, and up-to-date information increasingly must be bought at market rates. One solution already emerging is for governments in developing countries to make deals with small research companies, who see their developing country partners as less of a threat than the multinationals. The types of services involved could include development of specific products (contracts already exist for research in interferon, monoclonal antibody diagnostics, tea, coffee, cocoa and sugar-cane improvement), training of personnel, joint ventures and marketing arrangements.

In this situation, the information transfer networks being established by international agencies could have a crucial role. Many developing nations face similar problems and can benefit from knowledge and experience gained in other countries; it is also a means of bringing together the critical mass of research effort required to make real progress. Among the first initiatives were the Microbiological Resources Centres (MIRCEN), set up by Unesco to coordinate groups of researchers and to provide nodes through which information is transferred and contacts made. The International Agriculture Research Centres (IARC), created to improve food production in specific crops, have been very important institutions in developing and transferring research results to developing countries. They are well placed to expand their activities using biotechnologies and to build on their links with one another and with researchers in developed countries. The latest example of international cooperation is the International Center for Genetic Engineering and Biotechnology, founded under the auspices of UNIDO with centres in Delhi and Trieste.

There are a number of other United Nations organizations that have made investments in or studies of biotechnology, e.g. Unesco, FAO, WHO, ILO, UNDP and UNU. This widespread involvement is indicative of the breadth of the coming biorevolution, and the difficulty of tackling its implications by merely discussing, say, the farming system. For example, if World Health Organization research produced vaccines or pesticides effective against the vectors of yellow fever, malaria and trypanosomiasis, then African agriculture would be immeasurably changed. There are good reasons for all of these organizations to be involved.

Regional networks

Another method by which developing countries are attempting to overcome their late start and lack of a critical mass of researchers is to form regional networks. The UNDP and Unesco have provided funds for the countries of Argentina, Brazil, Chile, Costa Rica, Cuba, Mexico and Venezuela to organize a Latin American Biotechnology Network, which will be the first in the developing world. The project includes the eventual linking of the countries through a computer network, and also plans to contract with Latin American laboratories to produce biological research materials such as restriction enzymes, oligonucleotides, etc. The production of these materials not only provides important first-hand experience, but also loosens dependency on industrialized countries and demands for hard currency. The network will build on the experience gained from a Latin American biology training network.

CONCLUSION

Perhaps for the first time in history, developing countries are demanding a role in and a share of the benefits of a new technology before rather than after it has been deployed. To capture the promise of biotechnology will require a depth of commitment, honesty and realism that has sometimes been lacking among science policy-makers in developing countries, but — in contrast to technologies such as nuclear power, satellites, fibre optics and robotization — biotechnology offers areas of potentially valuable research to all but the smallest countries, and even these can participate in regional biotechnology networks.

Biotechnology will have unforeseeable economic impacts over the next twenty years and have far-reaching social repercussions. Agriculture, as much or more than any other human activity, will be transformed by 'biorevolution'. Biotechniques such as tissue culture can be practised even with small capital investments, yet can yield very real benefits to farmers. Plants that can expect little research attention from developed country scientists — such as tropical hardwoods — could provide handsome economic returns to a country willing to invest in research. Similarly, the development and use of microbial pesticides could contribute not only to diminishing the costs of imported pesticides, but would also be environmentally benign. There are so many technologies and so many possibilities in agriculture that choices of research goals will need to be made on the basis of rational analysis and not on speculative promises.

The increasing role of private industry in biotechnology means that it may become increasingly difficult for developing countries to secure unbiased advice from university scientists; the phenomenon of patenting and plant variety protection will also serve to block access to information produced in the industrialized countries. It is paradoxical that in this age of information and networking it is information and knowledge which will be increasingly withheld from the poorer countries even as these countries supply the essential genetic information in the form of germplasm. However, patent law is not always well enforced and the techniques of biotechnology once mastered are relatively simple to repeat, so that many of these products may be comparatively easy to produce.

Private capital sees the agricultural sectors of the developing economies as markets for the biotechnology products it is developing, such as pesticides and seeds; motives of profit maximization will determine the product developed and the purchasers will be those who can afford the price regardless of the impact on society. However, for developing countries, biotechnology is too important to allow narrow self-interest to control its deployment.

Subregional, regional and international cooperation also contributes to the promotion of research, exchange and training of biotechnologists, and to the dissemination of technical information. All these types of cooperation entail the adaptation of biotechnologies to the various economic and social environments.

References

BALTIMORE, D. 1982. Priorities in Biotechnology. In: National Research Council, *Priorities in Biotechnology Research for International Development*. Washington, D.C., National Academy Press.

BEARDSLEY, T. 1986. Cuban Biotechnology: Progress Despite Isolation. *Nature* (London), Vol. 320, No. 6057, p. 8.

BLACKWELL, J.H. 1980. Internationalism and Survival of Foot-and-Mouth Disease in Cattle and Food Products. *Journal of Dairy Science*, Vol. 63 (June), pp. 1019-30.

BONGA, J.; DURZAN, D. 1982. *Tissue Culture in Forestry*. The Hague, Martinus Nijhoff Dr. W. Junk Publishers.

BOONE, T.; MURCOK, D.; TALLEN, M.; MARTIN, F.; HOCKMAN, H.; ALTROCK, B.; DeOGNY, L.; LAI, P.; UYPYCH, J.; LANGLEY, K.; RUDMAN, C.; STEBBING, N.; SOUZA, L. 1983. Cloning and Expression of Chicken Growth Hormone in the Initial *E. Coli. DNA*, Vol. 2 (1), p. 74.

BRANTON, R.; BLAKE, J. 1983. A Lovely Clone of Coconuts. *New Scientist* (London), Vol. 98, No. 1359, pp. 554-7.

BUTENKO, R. 1984. Soviet Union: Decades of Progress. In: *Tissue Culture Technology and Development. Advance Technology Alert System (ATAS) Bulletin* (United Nations Centre for Science and Technology for Development, New York), Vol. 1 (November), pp. 31-5.

BUTTEL, F.; KENNEY, M.; KLOPPENBURG, J. 1983. From Green Revolution to Biorevolution: Some Observations on the Changing Technological Bases of Economic Transformation in the Third World. *Rural Sociology Bulletin*, No. 132 (August).

CANTLEY, M.; SARGEANT, K. 1981. Biotechnology: A Challenge to Europe. *Revue d'économie industrielle*, Vol. 18, pp. 323-34.

COCKING, E. 1960. A Method for the Isolation of Plant Protoplasts and Vacuoles. *Nature* (London), Vol. 187, pp. 927-9.

COHEN, S.; CHANG, A.; BOYER, H.; HELLING, R. 1973. Construction of Biologically Functional Bacterial Plasmids *in vitro*. *Proceedings of the National Academy of Sciences*, Vol. 70 (November), pp. 3240-4.

CONGER, B. 1983. *Cloning Agricultural Plants via* in vitro *Techniques*. Boca Raton, Florida, CRC Press.

DAVIES, R. 1981. Gene Transfer in Plants. *Nature* (London), Vol. 291, No. 5816, pp. 531-2.

EVANS, D.; BRAVO, J.; SHARP, W. 1983. Applications of Tissue Culture Technology to Development of Improved Crop Varieties. In: *Biotech 83: Proceedings of the International Conference on the Commercial Applications and Implications of Biotechnology*. Middlesex, Online Conferences Ltd.

FLINN, J.; JELLEMA, B.; ROBINSON, K. 1974. Barriers to Increased Food Production in the Lowland Humid Tropics. Ibadan, Nigeria. IITA memo.

GASCON, J.P. 1982. Les cultures tropicales *in vitro*. *Intertropiques*, pp. 21-5.

JAGANNATHAN, R. et al. 1978. Residual Effect of Blue-Green Algae Application on Rice Yield. *International Rice Research Newsletter*, Vol. 3, No. 4, pp. 20-1.

JAIN, H. 1982. Plant Breeders' Rights and Genetic Resources. *The Indian Journal of Genetics and Plant Breeding*, Vol. 42, No. 2, pp. 121-8.

KENNEY, M. forthcoming. *High Tech Biology: The Corporate Influence on » American Universities*. New Haven, Yale.

KOSUGE, T.; MEREDITH, C.; HOLLAENDER, A. 1983. *Genetic Engineering of Plants: An Agricultural Perspective*. New York, Plenum Press.

LIORET, U. 1982. Des palmiers éprouvette par millions. *La Recherche* (Paris), Vol. 13, No. 135, pp. 926-8.
MARIN, B. 1984. Les biotechnologies dans les pays en voie de développement pour quoi faire? *ORSTOM Actualités* (Paris), pp. 5, 6-7, 15
MCLOUGHLIN, P. 1970. *African Food Production Systems*. Baltimore, Johns Hopkins.
MOSHER, A. 1970. The Development Problems of Subsistence Farmers: A Preliminary Review. In: C. Wharton, Jr. (ed.), *Subsistence Agriculture and Economic Development*. London, Frank Cass.
NATIONAL BIOTECHNOLOGY BOARD OF INDIA. 1983. *Long Term Plan in Biotechnology for India*. New Delhi, Department of Science and Technology.
NATIONAL RESEARCH COUNCIL. 1984. *Genetic Engineering of Plants. Agricultural Research Opportunities and Policy Concerns*. Washington, D.C., Board on Agriculture, National Research Council, National Academy Press, 83 pp.
NORMAND-PLESSIER, F. 1984. Les biotechnologies au service de la sélection semencière. L'exemple du blé et du maWs. *Biofutur* (Paris), Vol. 24, pp. 23-30.
OFFICE OF TECHNOLOGY ASSESSMENT (OTA). 1984. *Commercial Biotechnology: An International Assessment*. Washington, D.C., United States Government Printing Office.
OKIGBO, B. 1982. Agriculture and Food Production in Tropical Agriculture. In: C. Christenson et al. (eds.), *The Developmental Effectiveness of Food Aid in Africa*. New York, Agricultural Development Council.
–. 1984. Present and Future Scope of Biotechnology at IITA. International Institute of Tropical Agriculture, Ibadan, Nigeria.
ORREGO, C. 1981. Evaluation of Microbial Technologies Involved in Fuel Production, Agriculture and Forestry. *World Bank Science and Technology Report Series*, No. 36 (August).
PEARSE, A. 1980. *Seeds of Plenty, Seeds of Want*. New York, Oxford.
PEEL, C.; BAUMAN, D.; GOREWIT, R.; SNIFFEN, C. 1981. Effect of Exogenous Growth Hormone on Lactational Performance in High Yielding Dairy Cows. *Journal of Nutrition*, Vol. 111 (September), pp. 1662-71.
POSTGATE, J. 1982. Biological Nitrogen Fixation: Fundamentals. *Phil. Trans. Royal Society of London*, B296, pp. 375-85.
QUINTERO, R. 1984. Framing a National Program on Biotechnology in Mexico. *ATAS Bulletin*, No. 1.
ROBERT, M.; LOYOLA, V. 1984. *Plant Tissue Culture*, *ATAS Bulletin*, No. 1.
ROCKEFELLER FOUNDATION AND OFFICE OF SCIENCE AND TECHNOLOGY POLICY. 1983. *Science for Agriculture*. New York, The Rockefeller Foundation.
RUTHENBERG, H. 1980. Farming Systems in the Tropics. Oxford.
SASSON, A. 1984. *Biotechnologies: Challenges and Promises*. Paris, Unesco Sextant Series, Vol. 2, 315 pp.
–. 1986a. *Quelles biotechnologies pour les pays en développement?* Paris, Unesco Biofutur.
–. 1986b. Recent and Foreseeable Progress in Plant Biotechnology and its Implications, Especially for the Developing Countries. In: *Life Sciences and Society*, Proceedings of an International Colloquium on Great and Recent Discoveries in the Biomedical and Social Sciences, and their Impact on the Evolution and Understanding of our Society, Stockholm, 26-8 November 1984, pp. 15-92. Amsterdam, Elsevier.
SCHULTZ, T. 1964. *Transforming Traditional Agriculture*. New Haven, Yale.
SCOWCROFT, W.; LARKIN, P. 1982. Somaclonal Variation: A New Option for Plant Improvement. In: I. Vasil; W. Scowcroft; K. Frey (eds.), *Plant Improvement and Somatic Cell Genetics*, pp. 158-79. New York, Academic Press.
SONDHAL, M.; SHARP, W.; EVANS, D. 1984. Biotechnology for Agriculture of Third World Countries. A paper prepared for the *ATAS Bulletin*.
SORJ, B.; WILKINSON, J. 1985. Modern Food Technology: Industrializing Nature. *International Social Science Journal* (Unesco, Paris) No. 105, pp. 301-13.
SPRAGUE, G.; ALEXANDER, D.; DUDLEY, J. 1980. Plant Breeding and Genetic Engineering: A Perspective. *BioScience*, Vol. 30 (January), pp. 17-21.
STEWARD F.; WIBBERLEY, G. 1980. Drug Innovation – What's Slowing it Down? *Nature* (London), Vol. 284 (13 March), pp. 18-120.

SWAMINATHAN, M.S. 1982. Biotechnology Research and Third World Agriculture. *Science* (Washington), Vol. 218, No. 4576, pp. 967-72.
TUDGE, C. 1984. Drugs and Dyes from Plant Cell Cultures. *New Scientist* (12 January), p. 25.
UBELL, R. 1982. Cuba's Great Leap. *Nature* (London), Vol. 302 (28 April), pp. 745-8.
UBELL, R. 1983. High-tech Medicine in the Caribbean. *The New England Journal of Medicine* (8 December), pp. 1468-72.
UYEN, N. 1984. The Use of Tissue Culture in Plant Breeding in Vietnam. Paper presented at the Inter-Center Seminar on IARCs and Biotechnology in Los Banos, Philippines, 23-27 April.
VIDALIE, H. (ed.). 1983. *La culture* in vitro *et ses applications horticoles*. Paris, Technique et Documentation.
YANCHINSKI, S. 1981. Bacteria to Textiles in U.K. Plant? *GEN* (March April), pp. 1, 3.
YANCHINSKI, S. 1985. Plant Engineered to Kill Insects. *New Scientist* (London), No. 1443, p. 25.

4

Health care technologies and health care delivery systems

Contributory factors and impacts

The 1978 Declaration of Alma-Ata affirmed that essential health care should be considered a basic human right and should be universally accessible at a cost that individuals and the community can afford (WHO, 1978). 'Essential health care' was defined broadly to include promotive, preventive, curative and rehabilitation services, as follows (Mahler, 1981):

— education concerning prevailing health problems and the methods of identifying, preventing and controlling them
— promotion of food supply and proper nutrition
— an adequate supply of safe water and basic sanitation
— maternal and child health, including family planning
— immunization against the major infectious diseases
— prevention and control of locally endemic disease
— appropriate treatment of common diseases and injuries
— provision of essential drugs.

It is immediately clear from this list that an enormous gulf separates the developed from the developing world. For most industrialized countries these problems belong to the distant past. Infant mortality and communicable diseases have been virtually eradicated. Adequate nutrition, clean water and access to medical treatment are taken for granted. Health problems arise more often from avoidable social behaviour (cigarette smoking, diet, alcoholism, drug abuse, traffic accidents) than from disease as such. Increasingly long-lived populations survive to suffer from cancers and dementias. Vast sums are spent on high technology curative care (microsurgery, transplants, scanners) mediated through specialist hospitals and personnel.

By contrast, for many developing countries these objectives seem utopian. They must face the much more basic, interrelated problems of communicable diseases, rapid population growth, low standards of living and poor nutrition. Development brings benefits, as the standard of living arises, but also creates further health problems. Resources are perforce limited, as regards both finance (some government health budgets are as low as US $4 per capita) and personnel; in spite of the very different problems, these scarce resources may be allocated to emulating inappropriate and costly Western models of health care rather than to adopting delivery systems suited to the particular circumstances of each society.

Efforts to improve the health care situation of the developing world are constantly undermined by a range of factors that favour the spread of disease, exacerbate the symptoms and hinder recovery.

RAPID POPULATION GROWTH

Between 1974, the year of the World Population Conference in Bucharest, and 1984, the year of the International Conference on Population in Mexico City, the world's population increased from 4,000 million to 4,700 million people. Unregulated fertility is strongly associated with such adverse health and social effects as early marriages, out-of-wedlock births, high incidence of maternal and infant mortality and morbidity, premature and low-weight infants, abortion, and family disruption. In the developing world between 10 and 20% of the children never live to see their first birthday, a total of 10 million infant deaths (Petros-Barvazian, 1984). Most of these deaths are avoidable, associated with or resulting from the poor health and nutritional status of mothers, and complications of pregnancy and childbirth.

Population policies should focus on delay of age at first pregnancy and increases in the intervals between pregnancies. Data from the World Fertility Survey for three Asian countries show that mothers under 16 years of age are twice as likely to lose their babies as those over 20. Mortality rates for children aged between one and two years are up to four times higher if their birth was followed by another within 18 months. The younger child born after a birth interval of less than two years also suffers, with higher perinatal and infant mortality rates. Children of birth-order seven or more have mortality rates one-third higher than those of birth-order two or three, with more frequent illness and slower growth; their mothers are older and physically exhausted; and large families tend to be poorer. Thus, women themselves suffer from unregulated fertility; maternal mortality rates range from 5 per 100,000 live births in Europe to up to 1,000 per 100,000 births in parts of Africa and Asia.

MALNUTRITION

All aspects of life and health, beginning with embryonic development and continuing on into old age, are intertwined with nutrition. Malnutrition may result from inadequate or excessive nutrient intake, as well as from nutrient imbalances and from host factors that interfere with proper absorption and utilization of nutrients. Inadequately nourished individuals may be stunted and wasted; their resistance to infection will be lowered; their work capacity will be reduced; and their productivity may be lowered. In addition, prolonged early undernutrition can contribute to impaired learning capacity.

The number of people affected worldwide by malnutrition is difficult to determine because of the many types of malnutrition and their differing severity. Estimates include the following (PATH, 1983):

— in developing countries, one quarter of the population (half a billion people) have energy-deficient diets

— about one-sixth of the world's children under the age of four (100 million children) have protein-energy malnutrition
— every year, half a million people go blind from vitamin A deficiency
— while the proportion of undernourished people is decreasing, the total number is increasing
— mild to moderate malnutrition, sometimes called marginal malnutrition, is up to ten times more prevalent than severe malnutrition.

Protein-energy deficiency is probably the major health problem of the world. Calorie and protein deficiencies in severe form appear as marasmus (starvation), kwashiorkor in children, or hunger edema in adults. Subclinical forms of protein-calorie undernutrition cause slowed growth, impaired development, greatly increased vulnerability to infections and lower productivity (Mata, 1975). Intestinal parasites can cause changes in the intestinal lining which exacerbates the nutritional deficiency: hookworms, for example, actually feed on the mucosa, causing bleeding and extensive damage. Chances of surviving diarrhoea, respiratory and other infectious diseases increase significantly as nutritional status of children is improved.

The most serious specific nutritional deficiency is iron deficiency owing to low intake of iron from predominantly vegetable diets, combined with intestinal blood loss from hookworm and schistosomiasis; anaemia is associated with increased mortality and morbidity, and reduced performance. Deficiency of vitamin A leads to blindness from keratomalacia in 50,000-100,000 children annually and less severe effects in many millions more (Crompton and Nesheim, 1984). Inadequate intake of vitamin A is prevalent among young children in developing countries where rice, cassava, white corn or potatoes are the dietary staples. Even slight deficits impair vision, reduce appetite, decrease growth, impair resistance to infection and interfere with reproduction in animals. Similar symptoms are observed in human populations known to consume inadequate quantities of vitamin A, but definitive studies of the effects of various levels of vitamin A deficiency in humans have not been made. Endemic goitre associated with iodine deficiency affects an estimated 200 million people. Other nutrient deficiency diseases include niacin deficiency (pellagra), thiamine deficiency (beriberi), vitamin D deficiency (rickets), and vitamin C deficiency (scurvy).

In addition, interactions of malnutrition with acute and chronic infections can be devastating, especially on brain development and mental performance (Winick, 1976). Infections can lead to loss of appetite, increased protein and energy expenditure, and metabolic loss of nitrogen, essential minerals and certain vitamins (Scrimshaw et al., 1968). Intestinal infections and periods of diarrhoea impair intestinal absorption of food, not only during the acute period, but also for some time thereafter, with persistent weight loss and poor growth rates in children. Collaborative research projects currently underway should lead to a greater understanding of the interrelationship between malnutrition and disease. A large longitudinal study is being conducted in Guatemala to determine whether mild-to-moderate protein-calorie malnutrition adversely affects the mental development of infants and preschool children (Barrett et al., 1982). Relationships among commercial agriculture, cash income, and health and nutritional status of children are under investigation in Kenya (Fleuret and Fleuret, 1983). The United States National Eye Institute is supporting research on measles and vitamin A deficiency among children under 5 years of age in India, aimed at preventing blindness (Underwood and Tupule, 1984).

Basic nutritional requirements often are proportionately greater for women and children, yet they tend to get the least food or the least protein of all family members; also, they are the first to suffer and the last to recover from temporary or chronic food shortages. The damage that malnutrition inflicts on pregnant and lactating women, infants and young children has profound effects both for them and for future generations.

Poor diets for pregnant women impair the health of both mothers and children. If they survive, low birth-weight infants tend to grow into small children and, upon reaching reproductive age, have a higher probability of producing low birth-weight infants (Hackman et al., 1983). This unhealthy cycle continues generation after generation. The provision of nutritional supplements to expectant mothers can reduce significantly the proportion of low birth-weight infants. For example, in Guatemala, supplement of only 20,000 calories during the entire pregnancy reduced the incidence of low birth-weight infants and improved the lactation performance of mothers (Klein et al., 1976). Both factors contribute to infant survival.

Abandonment of breastfeeding also has contributed to the increase in infant and child malnutrition in many developing countries. Various reasons for this phenomenon include changing work patterns of women during economic development, the perception of breastfeeding as being non-modern, and the strength of advertising campaigns by the infant formula companies. Substitutes for breast milk among the low-income population of developing countries tend to be expensive and nutritionally inappropriate, often contain infectious and toxic contaminants, and lack the non-nutritional factors found in human milk that are important in the development of host defences in an unsanitary environment. Furthermore, weaning foods are frequently inadequate in quality and quantity.

Improved maternal and child nutrition could help reduce infant mortality rates in developing countries from the present 50-250 per 1,000 live births to levels approaching those of the developed world (10-30 per 1,000 live births). Poor nutrition is closely related to inequitable distribution of physical resources and income in developing countries. Post-harvest losses in storage and transport diminish food supplies, while the shift toward cash crops has reduced the variety of foods formerly provided by subsistence farming. In addition, nutrition is influenced by multilateral and bilateral food assistance and trade policies, and by the marketing practices of multinational corporations.

UNSAFE WATER

Although globally there is enough fresh water to support current and anticipated populations on a sustainable basis, it is unequally distributed throughout the world and often it is unsafe to drink; in many societies the lack of safe water is the chief obstacle to meeting basic standards of health. As long ago as 1953, a report by UNICEF and the World Health Organization noted that 'probably three-fourths of the world's population drinks unsafe water, disposes of human excreta recklessly, prepares milk and food dangerously, are

Table 1. Principal diseases associated with unsafe water.

Disease (common name)	Persons infected (millions)	Controllable with clean water supply and basic sanitation (%)
Acquired in drinking or water contact		
Cholera	na	90
Typhoid fever	na	80
Diarrhoea	500[a]	50
Guinea worm	na	100
Schistosomiasis	200	10
Acquired in collecting water		
Malaria	300	na
Sleeping sickness[b]	na	80
River blindness	20-30	20
Elephantiasis[c]	270	na
Acquired by contact with excreta		
Roundworm	650	40
Whipworm	350	na
Hookworms	450	na

[a]Estimate is for annual cases in children in developing countries.
[b]Gambian trypanosomiasis.
[c]All filariasis infection.

Source: Chandler, 1984.

constantly exposed to insect and rodent enemies and live in unfit dwellings'. Unfortunately, despite some progress, such descriptions still apply today: more than half of the world's population has no reliable and safe water supply, and 70 to 80% has no sewage disposal (National Research Council, 1979). As a result, diarrhoeal diseases — the world's leading cause of infant mortality — are endemic throughout the developing countries. A sanitary water supply could eliminate half the diarrhoea, 90% of all cholera, 80% of all sleeping sickness, and 100% of Guinea worm infestation, as well as smaller fractions of other serious tropical diseases (US-AID, 1982) (see Table 1). Many people risk malaria, river blindness, or sleeping sickness because they must visit rivers and swampy areas to obtain water. Much of the poor hygiene in developing countries is due to the sheer difficulty of moving water, which weighs about 1 kg per litre, not counting the container. For adequate hygiene, people need about 20 litres per day; for each individual, this means carrying 20 kg of water each day. In a large family the burden is tremendous.

Geography or climate cause some of the problems, but in many instances

Table 2. Availability of clean drinking water and human waste disposal in selected countries.

		Share of population with service	
Country	Infant mortality (%)	Clean drinking water supply (%)	Human waste disposal (%)
Burkina Faso	21	31	na
Afghanistan	20	11	na
Angola	15	27	na
Ethiopia	15	16	14
Bolivia	13	37	24
India	12	42	20
Pakistan	12	34	6
Turkey	12	78	8
Indonesia	10	22	15
Tanzania	10	46	10
Honduras	9	44	20
Brazil	8	55	25
Mexico	5	57	28
Philippines	5	51	56
Chile	4.1	85	32
Costa Rica	2.7	72	97
Portugal	2.6	73	na
U.S.S.R.	2.6	76	na
Cuba	1.9	62	36
U.S.A.	1.2	99	99

Source: Chandler, 1984.

mismanagement of technology and abuse of fresh water sources are responsible. Fresh water can no longer be treated as a free good or as the principal means for disposing of human and industrial wastes.

Some observers have argued that water and sanitation systems should receive higher priority than other investments, including major reservoir projects, especially in view of the fact that some reservoirs have exacerbated the problems, for instance by tripling the incidence of schistosomiasis (Hunter et al., 1983). Clean drinking water, unfortunately, has not been a high priority for many countries (see Table 2).

MIGRATION AND URBANIZATION

Migration in developing countries traditionally involved movements of nomads, of the labour force, and of populations during disasters and wars, all of which pose significant problems but none as great as contemporary rural-to-urban movement of low-income populations stimulated by economic development and better employment opportunities and social services in the cities (World Bank, 1975b). Many cities in developing countries are growing at triple the rates of their national population increases, and some are reaching huge populations (see Table 3). Much of the urban growth is taking place without complementary expansion of urban services — housing, clean water, safe sewage disposal and regular garbage removal often are not available, facilitating the spread of parasitic and other infectious diseases. High population densities mean that more and more people are affected by natural or man-made disasters. Survivors often become refugees, who present enormous health and cross-cultural problems in many parts of the world. To cite one example, the flight of refugees from war and political disruption in South-East Asia has contributed, along with other movements of people, to the spread of falciparum malaria which is resistant to chloroquinine.

Table 3. Mega-cities in the year 2000 (population in millions).

Cities	1950	1975	2000
Mexico City	3.0	11.9	31.0
Sao Paolo	2.5	10.7	25.8
Tokyo	6.7	17.7	24.2
New York	12.3	19.8	22.8
Shanghai	5.8	11.6	22.7
Beijing	2.2	8.7	19.9
Rio de Janeiro	2.9	8.9	19.0

Source: Mukerjee, 1984.

Movement from one set of ecological conditions to another may expose people to particular diseases which are transmitted by insect vectors. Daily journeys to collect water or to collect firewood may result in contact with the blackfly and tsetse fly, the vectors of river blindness and sleeping sickness respectively. Seasonal movements for agricultural work take people away from permanent settlements, where dwellings have been protected with residual insecticide, to temporary dwellings which are unsprayed and harbour malarial mosquitoes. Disease itself may be responsible in turn for movements of people as whole areas become less habitable: river blindness and sleeping sickness together have brought about the desertion of fertile river valleys in northern Ghana, Ivory Coast and the south of Burkina Faso (Upper Volta).

STANDARD OF LIVING AND LIFESTYLES

Although difficult to document quantitatively, it is generally accepted that a society's standard of living is reflected in its health status. Poverty is the root cause of most health problems in developing countries: poverty is associated with undernutrition and malnutrition, and with high morbidity and mortality. Low income populations usually live in areas with contaminated water supplies and inadequate systems for disposal of human excreta and solid wastes. As incomes improve, so do housing, sanitation and water quality, reducing exposure to and transmission of vector-borne and water-borne diseases. Poverty also limits access to health care, including preventive services — even where health services are provided free of charge, the costs of transportation to a health care facility and loss of work time often pose major barriers to the very poor. In one study, the decline in infant mortality between 1950 and 1970 in Ghana was causally associated with general improvements in the standard of living during that time period, since health services had scarcely reached the majority of the population (Gaise, 1979). Development economists are increasingly recognizing that the quality of labour input is affected by the health of the working population. Efforts to collect and analyse data on economic benefits resulting from health investments tend to show that improvements in health status are associated with increased productivity (Griffith et al., 1971).

Better living standards are not always an unmitigated benefit, however. The two major causes of death after the age of five years in industrialized countries, cardiovascular disease and cancers, are associated with more affluent lifestyles (in particular diet and cigarette smoking) that developing countries often willingly adopt as their standard of living improves — consequently their populations are becoming prone to the same problems. Changes of diet towards more refined and sweetened foods and smoking also exacerbate dental disease. In addition, accidental injury or death as a result of work-place or vehicular accidents tend to increase with economic development: statistics are far from comprehensive, but some idea of the magnitude of the problem is given by the International Labour Organization estimate (for only 46 countries) that 9 million people were injured in 1982 as a result of on-the-job accidents, 24,000 fatally (*World Health*, 1984a).

It is alarming that, while smoking is to some extent being brought under control in developed countries, developing nations are increasingly becoming targets for highly sophisticated campaigns promoting smoking. In 1978, the WHO Expert Committee on Smoking Control warned that: 'In the absence of strong and resolute government action, we face the serious probability that the smoking epidemic will have affected the developing world within a decade and that...failing immediate action, smoking diseases will appear in developing countries before communicable diseases and malnutrition have been controlled and the gap between rich and poor countries will thus be further expanded' (WHO, 1979b). The 1982 WHO Expert Committee on Smoking Control regretfully acknowledged that the previous predictions were coming true (WHO, 1983d).

There is no standard form of international data collection and data reporting on smoking. In almost all developing countries for which data are available, however, 50% or more of adult men are dependent on some form of tobacco use, mostly smoking. In the People's Republic of China about 25% of the men are addicted to tobacco smoking before they reach the age of 18 years, and in India roughly a third are addicted before reaching the age of 20 years, with considerable regional variation. Smoking is prevalent among schoolchildren, particularly boys. By contrast, less than 5% of women are smokers, but the rates could increase in imitation of women in industrialized countries.

In view of the persistent shortage of resources available for health care, the additional burden of treating smoking-related diseases could have dire consequences for developing countries. Any short-term economic advantages of growing tobacco and manufacturing tobacco products will be outweighed by mortality and morbidity due to smoking. Developing countries have a unique opportunity to take firm steps. Failure to take action to halt the most avoidable of modern epidemics will hinder achievement of the goal of Health for All by the year 2000. Action to prevent smoking should be aimed at all people, but especially children and young adults.

EFFECTS OF DEVELOPMENT ON ECOLOGY AND HEALTH

Economic development and technological advances have both positive and negative effects on ecology and health.

Agricultural development

Advances in agricultural technology have brought an improved standard of living to many people in developing countries and with it improved health, but at the same time some unanticipated side-effects have had a negative impact. One example is the increased incidence and severity of schistosomiasis which has accompanied construction of dams and irrigation systems (Warren, 1980). Construction of hydroelectric dams and other human activities in river basins have been associated with redistribution and increase of tsetse flies (carriers of human and bovine trypanosomiasis) and of simulin flies (carriers of onchocerciasis and leishmaniasis). In addition, river basin development is associated with large-scale population migrations, accompanied by increases in morbidity and mortality rates (Scudder, 1972).

Desertification results from a combination of unsound farming and grazing practices, rapid growth of human and livestock populations, inadequate and uncoordinated development policies, and climatic fluctuations with serious consequences in terms of food supply, nutrition, disease and income (Brink et al., 1977). Approximately 50 million people are now living in areas that are slowly turning to desert. Other types of land degradation include encroachment of savanna woodlands and erosion of mountain habitats in such countries as Peru and Nepal. Changes in agricultural practices from mixed cropping to monocropping for cash and export affect the local ecology and may result in nutritional disorders and increased incidence of parasitic and other infectious diseases.

Another major concern has been the dramatic increase in the use of pesticides: in some countries, more than 1,000 compounds may be registered as pesticides, formulated in many thousands of commercial products. Their overuse has become an alarming problem in developing countries, since few farmers try to buy and apply pesticides according to recommendations of state agronomists, relying instead on information from neighbours, local retail stores, or pesticides salesmen and the general principle that 'if some is good, surely more is better'. One result is a high incidence of human poisonings: although reliable mortality and morbidity data on pesticide poisonings are not available, it has been estimated that there are 375,000 cases of human poisonings by pesticides every year in developing countries, with some 10,000 deaths. The contamination of wheat and flour with highly toxic pesticides during transport has caused epidemic outbreaks of poisoning in Brazil, Colombia, Jamaica and Mexico (Waldemar, 1984). Many of the pesticides used in developing countries have been banned from production and or use in the developed world because of high toxicity, but are still exported to or produced in the developing countries (Navarro, 1984).

The indiscriminate and unsafe use of pesticides has also been responsible for development of resistance in agricultural pests and for increase in total number of pest species because natural predators have been exterminated by pesticides. For example, in Central America, the number of major cotton pests rose from 2 to 7 in the first 10 years of pesticide use; in desperation the number of pesticide applications was increased from 2 to 20 per season. Overuse of pesticides on crops like cotton or tobacco or rice also reflects the power of large landowners and the conflict between rich and poor in many developing countries.

Industrial development

Technological advances in industry — as in agriculture — result in a higher standard of living and improved health, but they may be accompanied by extreme stresses on the environment in terms of toxic substances and hazardous wastes. Many biological scientists and engineers believe, however, that they can utilize traditional and newly emerging techniques to make substantial progress in controlling environmental pollution and do so safely, with critical evaluation of potential consequences, for example by developing the natural biodegradative capabilities of certain organisms using modern genetics.

The best strategy for efficacy and safety is to isolate organisms from environments of interest — sediments, sludges, dumps, soils — introduce a plasmid or gene coding for the biodegradative enzyme(s), and then return the organism to its habitat. Potential ecological hazards should be minimal so long as genes are transferred within bacteria or within fungi, as opposed to moving mammalian genes into plants or bacteria (Omenn and Hollaender, 1984).

Like the pesticides mentioned above, hazardous substances banned from production or use (or highly regulated) in the industrialized nations may not be controlled in developing countries. Asbestos products which require warning labels in Australia, the United States of America or the Federal Republic of Germany are sold in Indonesia, India, Africa and the Middle East without such warnings. After the Federal Republic of Germany established more stringent safety and health regulations, Spain, Yugoslavia and the Republic of Korea became the primary producers of asbestos-related products sold in the Federal Republic of Germany (Navarro, 1984). Similar experiences have been reported for the regulated carcinogenic dye benzidine, for arsenic and refined copper from primary smelters, mercury mining, lead smelters and battery plants, and other industries.

The United Nations General Assembly in its 34th session in 1979 adopted a resolution urging 'member states to exchange information on hazardous chemicals...that have been banned in their territory and to discourage, in consultation with importing countries, the exportation of such products to other countries'. But little has been done. Sweden, for example, only instructs its corporations to abide by the laws and practices of the importing or recipient country. European Economic Community directives dealing with classification, packaging and labelling of dangerous substances (67 548 EEC) and banning the use in EEC countries of certain pesticides and other substances (79 117 EEC) do not include any provisions regarding exportation of these substances. Concerns about consequences for developing countries and their workers and demands for international protection run into criticisms that such views are 'elitist', reflecting problems of the wealthy, requiring pollution control actions that poor countries cannot afford. According to the *World Environment Handbook* (1984), 110 developing countries now have environmental agencies, compared with 11 agencies in 1972, but many countries lack the financial and human resources to enforce their environmental laws, guidelines and regulations.

Finally, given the shortage of suitable sites and resistance to incineration of hazardous wastes in developed countries, there have been efforts to identify disposal sites in developing countries for toxic wastes.

Measures for improved health

EPIDEMIOLOGICAL SITUATION

Communicable diseases

Infections are the principal cause of morbidity and mortality in young infants and children in the developing world: in certain countries up to 40% of children die before five years of age (National Academy of Sciences, 1977). The major diseases of the developing world, their extent and distribution, characteristics and control measures are set out in Tables 4 to 6.

Poor living conditions and malnutrition exacerbate many of the infections, and create difficulties in diagnosing and treating disease. Improvement in the standard of living is widely agreed to be the main factor in controlling tuberculosis; better hygiene and nutrition would help to reduce the incidence and severity of diarrhoea, tetanus, measles, polio and hepatitis. Although great strides have been made in immunization, chemotherapy and vector control, many of the advances have not yet reached societies in greatest need.

Effective treatments exist for the major bacterial infections, apart from leprosy. The treatment of *viral* diseases is less straightforward, since they are often complicated by secondary bacterial infections (WHO, 1981, 1983c; Rockefeller Foundation, 1984), and they are not controllable by drugs. Immunization offers the best hopes, though there have been problems with

Table 4. Major bacterial and mycobacterial infections.

	Morbidity (cases/yr)	Mortality (deaths/yr)	Geographic distribution	Disease characteristics	Control measures	Research priorities
Diarrhoea	3-5 billion[a]	5-10 million[a]	Global	Fever, acute dehydration	Improved sanitation Early administration of oral rehydration therapy (ORT)	Improved oral rehydration programmes Vaccine development Field epidemiology and clinical studies
Pertussis	20 million[a]	735,000[b]	Global	Cough, lung involvement, slow recovery	Immunization	Improved administration of immunization programmes
Tetanus (and neonatal tetanus)	150,000[a] (neonatal)	1.1 million[b]	Global	Painful muscular contractions and spasms; breathing obstruction	Immunization Sanitary birthing conditions	Improved administration of immunization programmes
Diphtheria	800,000[a]	55,000[a]	Global	Fever, severe throat heart and nervous system involvement	Immunization	Improved administration of immunization programmes
Tuberculosis	10 million new cases[c]	3 million[c]	Global	Infection of lungs and other organs with fevers, systemic toxicity	Improved sanitation Improved standard of living Case-finding Surveillance Mass BCG immunization	Pathogenesis of the disease Genetics of mycobacteria Improvements in immunization Short-term chemotherapy Sociological studies
Leprosy	Prevalence: 12 million[a]	Low	Primarily tropics	Incubation period 3-10 years or longer Lepromatous form severe; tuberculoid form benign	Case-finding Early treatment	Improved chemotherapy Vaccine development and field testing

[a]Walsh and Warren, 1979.
[b]Estimated in developing countries (excluding China) for 1983 (Rockefeller Foundation, 1984).
[c]WHO, 1982a.

Table 5. Major viral infections.

	Morbidity (cases/yr)	Mortality (deaths/yr)	Geographic distribution	Disease characteristics	Control measures	Research priorities
Measles	80 million[a]	2.5 million[b]	Global	Fever, rash; pneumonia, blindness, deafness	Immunization Improved nutrition	Relationship between malnutrition and vaccine effectiveness Improved immunization programmes
Poliomyelitis	2 million[a]	15,000[a]	Global	Fever, malaise, paralysis	Immunization Improved sanitary conditions	Efficacity of vaccine under varying field conditions Improved nutrition programmes
Rotavirus infantile diarrhoea	High	1 million[c]	Global	Fever, acute dehydration	Improved sanitary conditions	Etiology Vaccine development
Hepatitis B	200 million carriers[d]	High	Endemic in Asia, Africa, Latin America	Acute illness of variable severity	Improved sanitary conditions Improved personal hygiene Screening of blood donors Administration of immuno-globulin	Vaccine for active immunization Field testing of vaccines available Relationship between hepatitis B and liver cancer
Respiratory diseases including influenza	Very high	4.5 million[a]	Global	Upper and lower respiratory involvement, fever	Vaccines for some specific viruses Improved nutrition	Vaccine development Antiviral drugs Improved clinical care Improved surveillance
Arboviral diseases:						
Dengue fever	1.5 million[a]	100[a]	Endemic in tropical Asia, certain Pacific Islands, South & Central America, Caribbean	Fever, rash, encephalitis	Control of vectors	Pathogenesis Vaccine development Anti-viral drugs
Japanese encephalitis	High	High	Tropics and subtropics; Thailand, Burma, India	Fever, rash, encephalitis	Control of vectors Vaccinations	More potent inactivated cell-culture vaccine

Walsh and Warren, 1979.
Estimated in developing countries (excluding China) for 1983. (Rockefeller Foundation, 1984).
Vesikari et al., 1983.
WHO, 1983c.

Table 6. Major tropical parasitic diseases.

	Persons infected/yr[a]	Morbidity (cases/yr)[a]	Mortality (deaths/yr)[a]	Geographic distribution	Specific parasite (source for diagnosis)	Control measures	Research priorities
Malaria	800 million	150 million (fever, coma)	1.2 million	Global	Protozoa, *Plasmomodium* species(blood)	Treat infected persons and eliminate the parasite; attack the vector (female *Anopheles*) in its larval or adult stages or prevent from breeding; prophylaxis for susceptible persons; impede man-vector contact (mosquito nets, screens, repellents)	Biology of the parasite; isolate surface antigens to make vaccines; drug development; field research
Schistosomiasis	200 million	20 million (liver and urinary tract fibrosis	0.5-1 million	N. & C. Africa Brazil China	Blood flukes (trematodes) *Schistosoma* species (feces)	Reduce contamination of natural water by human excreta; attack snails with chemical molluscicides combined with correct design of irrigation systems; reduce human contact by providing clean, potable water supplies and safe water for recreation	Immunology; chemotherapy; vector biology and control
Filariasis (lymphatic)	250 million	2-3 million (elephantiasis)	low	Tropical Africa Mexico Guatemala	Filarial worms *Wucheria bancrofti*, *Brugi malayi* (blood)		

(onchocerciasis)	30 million	0.2-0.5 million (blindness)	20,000-50,000		*Onchocerca volvulus* (skin snips)	Control vector anthropods to interrupt transmission; eliminate parasite reservoir in man by mass chemotherapy with DEC; prevent infection by environmental methods, personal hygiene, or chemo-prophylactic drugs	Better understanding of natural history, epidemiology, and vectors; improve use of existing filaricides and find new ones; identify filarial antigens for serodiagnostic texts and the development of vaccines
African trypanosomiasis	1 million	10,000 (sleeping sickness)	5,000	Tropical Africa	Ptotozoans *Trypanosoma gambiense* (chronic) (blood) *Trypanosoma rhodesiense* (acute) (blood)	Reduce reservoir in man; Tsetse fly control; medical surveillance to find early cases	Epidemiology and control; chemotherapy and drug development; immunology
South American trypanosomiasis (Chagas' disease)	12million	1.2 million (heart disease)	60,000	Brazil Venezuela	*Trypanosoma cruzi* (kissing bug)	Residual insecticides; improved housing and sanitary conditions (removing places where triatatomid bugs ('kissing bugs') live and breed)	Epidemiology of the disease and studies of the vectors; chemotherapy and drug development; immunology
Leishmaniasis	12 million	12 million (kala-azar, skin sores)	5,000	Africa India	Various parasites of the genus *Leishmania* (splenic aspiration)	Reduce man-vector contact by mechanical means (bed-nets, screens, repellents); reduce sandfly populations with insecticides; control animal reservoirs	Increase knowledge of the geographical distribution and varieties of the disease; epidemiology, parasitology, vector biology, disease control; typing of sandflies and of leishmania; drug development and field testing.

[a]Adapted from Warren and Bowers, 1983.

the development of vaccines for some viruses. In the case of polio, the live trivalent vaccines that have been very successful in many countries seem to encounter difficulties in certain tropical and subtropical countries; however, highly purified, potent inactivated polio vaccines (Salk approach) are now becoming available.

Global estimates of the prevalence of *parasitic* diseases are staggering and increasing. The total number of parasitic infections far outnumbers the world population since multiple infections are the rule, not the exception, in most tropical areas; estimates of infection rates reflect the rates of diagnosis, which are limited by the sensitivity of the techniques used. Tropical parasitic diseases affect every aspect of human life from childhood onwards; among survivors they can disable an entire population. Parasitic diseases often are not amenable to simple control measures, because they are intimately associated with human behavioural factors, agricultural practices, domestic and work animals, and general aspects of environment, sanitation, poverty and socioeconomic structures. In addition, many parasitic diseases have complicated transmission cycles, including reservoirs in the animal world (see Warren and Bowers, 1983).

Traditional methods of controlling major parasitic diseases through a combination of vector control and chemotherapy are encountering problems as insects and organisms develop resistance to insecticides and drugs. In addition, the application of these techniques can be both costly and cumbersome. The single largest recent problem in malaria control, for example, is resistance of mosquitoes to insecticides: at least 51 species of anopheline mosquitoes have become resistant to one or more insecticides (34 to DDT, 47 to dieldrin and 30 to both DDT and dieldrin) (WHO, 1980b). Several vector control techniques are under development or consideration. Alternative insecticides are the most immediate answer; however, of 2,000 compounds tested since 1960, only 10 have been suitable for use as residual insecticides. Rotation or mixtures of insecticides and genetic manipulations aimed at reducing vector populations have not proved successful. Eventually, integration of chemical with biological and or environmental methods may reduce the present dependence on chemicals and overcome vector resistance. Environmental manipulation to prevent or reduce vector breeding through the alteration of habitats and breeding habits has yielded effective vector control; despite high initial costs, the benefits are often long-term and cost-effective.

There have been striking advances over the last 15 years in the treatment of schistosomiasis with chemotherapy, particularly the development of short-term drug regimens (WHO, 1980a). Chemotherapy has been less successful, however, in the treatment of some other parasitic diseases because of adverse side-effects (e.g. filariasis, and especially trypanosomiasis, where the treatment itself has a fatality rate of 5-10%). Drugs are active against Chagas' disease only in the early stages of infection, which usually occurs in childhood although the disabling or fatal effects may not become apparent until 10-20 years later (Brenner, 1982).

Coronary heart disease

Cardiovascular diseases leading to heart disease and stroke remain the leading cause of mortality and morbidity in industrialized countries, despite substantial knowledge and impressive progress in prevention and control. Atherosclerosis causes fibrous fatty changes, proliferation of the endothelial lining and obstruction of the arteries serving the heart muscle, leading to sudden death, myocardial infarction, angina pectoris, congestive heart failure, or arrhythmias. Since population rates for coronary heart disease have increased substantially over relatively short periods of time, and since rates in migrant populations (Japanese in Hawaii, Irish in Boston) tend to increase to match those of their adopted cultures, powerful environmental factors must be involved. Smoking, obesity, sedentary living and excessive dietary saturated fats must be avoided to prevent the development of high-risk-factor patterns. Coronary disease is already a serious burden in some developing countries and will be a threat to many more as socioeconomic development progresses. A comprehensive plan for prevention of heart disease should combine (1) a population strategy to alter lifestyles and their social and economic determinants; (2) a high-risk strategy for bringing preventive care to individuals at special risk, and (3) secondary prevention aimed at averting recurrences in those already affected (WHO, 1982b).

Cancers

There is little awareness of the potential magnitude of the cancer problem in developing countries among the general public or among health administrators, even in those countries where cancers are overtaking infectious diseases as a major cause of death. There are already more cancer deaths in the developing countries (2.3 million) than in the developed regions (2.0 million). A rough WHO estimate, based on expected demographic and health status trends, suggests that the number of cancer deaths may rise by more than 50%, to approximately 8 million annually, by the year 2000 (WHO, 1979a).

Globally, the three most frequent cancers are stomach cancer, with some 680,000 new cases yearly; lung cancer, with 590,000; and breast cancer with 540,000 (Parkin et al., 1984). Cancers that chiefly affect the developing world are those of the uterine cervix, with an estimated 460,000 new cases yearly; oesophagus, 300,000; liver, 260,000; and mouth, 100,000. Lung cancer will become epidemic in developing countries unless the current increase in the consumption of cigarettes is slowed or reversed (WHO, 1979b, 1984d).

Primary liver cancer, which is among the most common cancers in South-East Asia, the Western Pacific, and Africa south of the Sahara, is of particular importance because a preventive strategy is now available. Epidemiological investigations indicate that there is a consistent and specific causal association between hepatitis B virus and hepatocellular carcinoma, accounting for 80% of such cancers (see p. 175). Hepatitis B virus is thus second only to tobacco (cigarette smoking) among the known human carcinogens.

Mouth cancer is the commonest form of cancer in South-East Asia, where one-sixth of the world's people live; 90% of cases are caused by local forms of betel nut and tobacco quid chewing and smoking. Up to 15 years may elapse before lesions in the mouth (called leukoplakia) turn cancerous. If the lesions are detected in sufficient time for treatment, the disease is curable through radiotherapy and surgery. Unfortunately, most persons seek help only when they are in pain, which is a late advanced symptom, when the cancer is usually incurable. Chemoprevention with vitamin A may be especially attractive in populations where vitamin A deficiencies need to be corrected anyway (Prentice et al., 1984).

Expensive therapies are unrealistic in most developing countries, but the wider application of our current knowledge of cancers and of cancer prevention could reap tremendous benefits. There is the knowledge to prevent a third of all existing cancers and to cure a third if cases are detected early enough (*World Health*, 1984b).

Oral diseases

Oral and dental diseases are causes of much pain and disability. In some parts of the developing world, including most of Latin America, virtually every person develops dental cavities. The problem has increased steadily with modernization and consumption of refined foods. Control of caries can, of course, be accomplished with fluoride.

Chronic periodontal diseases are found in all societies, but these destructive diseases of the tooth-supporting tissues are especially prevalent and severe in Asia and Africa. Poor oral hygiene, low socioeconomic status, frequent pregnancy, poor nutrition, smoking, and betel nut and khat chewing are associated with periodontal disease. Because of the reversibility of these factors, periodontal disease is a preventable chronic disease (Carlos, 1973).

Mental illnesses

Mental health problems are probably at least as serious in the less developed countries as they are in the more advanced countries, but accurate statistics are unavailable. Worldwide, at least 40 million people in the world suffer from severe forms of mental disorder, such as schizophrenias and dementias; some 20 million suffer from epilepsy; and a further 200 million are incapacitated by less grave mental and neurological conditions, such as severe neuroses, mental retardation and peripheral neuropathies (WHO, 1984a). Alcohol- and drug-related problems and mental disorders secondary to physical disease add to the problem. Few people in the developing world have access to mental health care. Most people suffering from epilepsy receive no treatment, despite the fact that treatment is usually cheap, simple and effective. Assistance in preventing and treating mental illnesses and in training health professionals is complicated by cultural and social barriers.

GENERAL MEASURES OF ENVIRONMENTAL HYGIENE

The health problems associated with unsafe water supplies have already been outlined; the solution lies in improved waste disposal, but high cost and cultural barriers may hinder sanitation development. The World Bank estimates that indoor water and sanitation for all the developing countries would cost US $800 billion to construct and US $10 billion per year to operate and maintain (Kalbermatten et al., 1982). Consequently, only inexpensive water supply systems and excrement disposal systems will be practical in the developing countries. Fortunately, several appropriate technologies are available and affordable (see Table 7).

Table 7. Cost of minimal levels of human waste disposal.

Low-cost toilet	Cost per household: Total Investment	Cost per household: Hypothetical total per month	Share of income of average low-income household
	(1983 US dollars)	(1983 US dollars)	%
Poor-flush toilet	70	2.0	2
Pit latrine	125	2.6	3
Communal toilet	355	8.3	9
Vacuum-truck cartage	105	3.8	4
Vietnamese toilet	50	—	—

Source: Chandler, 1984.

The design of water supply systems will vary according to local conditions. Surface water supply, cisterns, or wells may be required. Large drill rigs are expensive to buy and operate and difficult to transport over poor roads, but small, jeep-mounted rigs are capable of drilling to a depth of 30 m. In order to reach a water table, pipes can be hammered as much as 15 m into the ground by a person using a sledgehammer: the system requires a pump; it must be inexpensive, reliable and easy to repair locally; and it must be covered to prevent contamination by humans or animals (Bourne, 1984).

The aqua-privy has been widely installed in India, largely through the effort of individuals and of local entrepreneurs (Pathak, 1982). It is a single tank, partly filled with water, into which human waste is carried via a drop-pipe. The system is 'sealed' by a layer of scum on the water surface that keeps odours in, allowing breakdown of waste without oxygen. The drop-pipe extends below the water line from either a squatting plate or a pedestal that can be flushed with two or three litres of water. An overflow outlet, also submerged, carries effluent underground, so the toilet must be well above the water table. Digested materials can be removed periodically by the owner.

One inexpensive system consists of an intake that separates urine and faeces plus two above-ground concrete chambers, used alternately, that anaerobically digest human waste. Because the system is sealed, it can be used in flood-prone or high water table areas. Pouring ashes into the chambers promotes decomposition and makes the waste suitable as a fertilizer within a year. The separated urine travels to an ash-filled container and remains for a few days, after which it too can be used as fertilizer. This system is reportedly used widely in the Socialist Republic of Viet Nam, and it has been introduced in the United Republic of Tanzania, Mozambique and elsewhere (Environmental Sanitation Information Center, 1983).

Less sophisticated systems, including the cesspool used widely in Thailand and the simple pit privy used around the globe, can create serious problems. Expertise is needed to instal the systems, since they carry considerable risk of failure from flooding. Projects seem to work best when people are taught to construct and operate their own systems. Primary health care workers can make sure the facilities are used properly, but developing country governments are often organized so that sanitation projects are administered by an agency other than the Ministry of Health.

VACCINES

Currently available priority vaccines

Immunization is our most potent and cost-effective strategy to fight specific diseases yet, even for diseases fully preventable with vaccines, immunization services remain tragically underutilized in the world today. In 1983, an estimated 5 million children died in developing countries and another 5 million were crippled, blinded or otherwise disabled from the six diseases targeted by the WHO Expanded Programme on Immunization (see Table 8), despite the fact that these vaccines and the means to deliver them are readily available and inexpensive. Costs in established programmes average only US $5.00 per fully immunized child, although costs may be higher during programme development. Since infants comprise 4% or less of the population, a national immunization programme can be implemented for US $20 or even as little as $0.60 per capita.

The World Health Organization established the Expanded Programme on Immunization (WHO-EPI) in 1974, with the goal of providing immunization for all the children of the world by 1990 against measles, pertussis (whooping cough), tetanus, poliomyelitis, diphtheria and tuberculosis. This will require planning, management, appropriate technological modifications for field conditions, and an effective health infrastructure. Vaccines must be ordered in advance in appropriate quantities, supplied regularly to health units, and kept potent in an unbroken 'cold chain' at safe temperatures from the place of manufacture to the place of use. There must be an infrastructure of service delivery points and field workers to make immunization accessible in villages and rural areas, as well as in cities, and to help people understand the benefits of immunization and use the services offered.

Although much has been accomplished since 1974, acceleration of the programme is required if the 1990 goal is to be attained (see Tables 8 and 9). In 1983, only 30% of children under the age of one year had completed the three-dose immunization with DPT vaccine, and a still smaller proportion had received polio or measles immunization. Major difficulties are insufficient political will to provide the needed financial and human resources, insufficient management capacity to translate available resources into effective programmes, and instability of measles and polio vaccines under field conditions. In addition, many children are denied immunizations because of overcautious lists of contraindications (Galazka et al., 1984).

The WHO-EPI did not formulate a universal set of recommendations for immunization of children, because WHO believes each country should formulate its own policies, but its principal recommendations can serve as a general guide: (1) health workers should use every opportunity to immunize eligible children, giving BCG, DPT, oral polio and measles vaccine in the first year; (2) the risks of serious complications from these vaccines are much lower than the risks from the natural diseases; (3) children with malnutrition, low-grade fever, mild respiratory infections, diarrhoea or other minor illnesses should not be denied immunization; and (4) hospitalized children should receive appropriate immunization before discharge.

In addition to the vaccines included in the WHO programme, immunization has proved effective against several mosquito-borne and tick-borne encephalitic viruses (notably yellow fever, for which a successful attenuated live vaccine has existed since the 1940s).

Table 8. Estimated immunization coverage with DPT, poliomyelitis, measles, BCG and tetanus vaccines in developing countries (excluding China), ranked by number of infant births, 1982.

Country	Total no. of newborns (millions)	Cumulative % of births	Immunization coverage (%)				
			Children under 1 year of age				Pregnant women
			DPT III	Polio III	Measles	BCG	Tetanus
India	24.84	27	39	18	0	18	24
Indonesia	5.30	33	29*	3	2	55	15
Brazil	4.41	38	53	99+	64	61	...
Bangladesh	4.21	42	2	2	2	3	1
Pakistan	4.11	47	4	4	6	1	1
Nigeria	4.07	51	...	...	...	...	...
Mexico	2.93	54	23	73	8	25	...
Soc. Republic of Viet Nam	2.18	57	...	...	...	...	...
Philippines	1.87	59	51*	44	...	61	...
Thailand	1.78	60	53*	33	0	73	30
Islamic Republic of Iran	1.69	62	35	62	46	8	3
Turkey	1.60	64	64	69	52	47	...
Ethiopia	1.54	66	6	6	7	10	...
Egypt	1.51	67	...	...	...	...	...
Burma	1.37	69	9	1	0	19	10
Zaire	1.34	70	18	18	20	34	...
South Africa	1.09	71	...	...	...	...	...
Morocco	0.98	73	44	44	...	...	...
Rep. of Korea	0.96	74	61	62	5	99+	...
Algeria	0.95	75	33	30	17	59	...
Tanzania	0.92	76	58	56	82	84	35
Sudan	0.89	77	2	4	4	2	1
Kenya	0.88	78	...	...	...	...	...
Colombia	0.85	78	21	22	22	53	...
Afghanistan	0.70	79	5	5	8	10	1
Total	72.97		28	24	9	24	11
All other developing countries	19.20	21	42	23	32	31	10
Grand Total	92.17	100	31	24	14	26	11

**Coverage for DPT II is given for those countries using a 2-dose immunization schedule.*

... Data not available to WHO/Geneva

Source: Rockefeller Foundation, 1984.

Table 9. Percentage of countries or areas known to be involved in selected immunization activities, by WHO Region, as of 31 December 1983.

	Africa	Americas	South-east Asia	Europe	Eastern Mediter-ranean	Western Pacific
Number of countries or areas in the region	46	47	11	37	24	32
Vaccines included in national immunization schedules						
BCG	100	92	100	73	100	94
DPT	100	100	100	92	100	100
Measles	96	96	64	95	100	94
Poliomyelitis	98	100	91	100	100	100
Tetanus for women of childbearing age	96	96	82	3	100	94
Reporting to WHO						
Total number of immunizations	59	100	91	3	92	88
Immunization by age or dose	22	62	91	3	92	66
Reporting to WHO the 1983 incidence of						
Diphtheria	52	68	82	8	66	94
Measles	72	87	91	8	66	88
Pertussis	61	70	91	11	66	91
Poliomyelitis	52	68	91	8	66	84
Tetanus	48	68	91	8	66	91
Tuberculosis	65	2	82	5	70	88
All of the above	33	2	82	3	66	75
Neonatal tetanus	20	47	46	3	38	3
Coverage surveys	61	15	64	11	50	34
Staff participating in						
EPI planning & management course	89	64	82	22	71	72
EPI mid-level management course	72	64	91	5	24	31
EPI cold-chain course	28	45	91	3	32	25
Organization of EPI national mid-level management course	63	83	82	5	21	47
Incorporating EPI training materials in national training curricula	22	15	82	11	12	25
Programme reviews	37	32	46	3	38	12

Source: Rockefeller Foundation, 1984.

New vaccines: work in progress

Malaria

The Sixth Report on the World Health Situation 1973-7 described malaria as the greatest killer among the tropical diseases (WHO, 1980c). After the age of 12 months, almost every child in tropical Africa has contracted malaria; at least one million children die of the disease every year. In spite of eradication programmes based on the destruction of mosquitoes with insecticides and on chemotherapy, there has been a recent resurgence of the disease, especially in South-East Asia and Latin America (Chapin and Wasserman, 1981).

Through rapid advances in molecular genetics and immunology, vaccines are being developed against each stage in the life-cycle of the malaria parasite: sporozoites, which develop in mosquitoes and are injected by them into humans; asexual erythrocytic parasites, which cause the disease; and sexual stages, which develop in humans and transmit the infection to mosquitoes. An effective vaccine against sporozoites could enable the human immune system to kill sporozoites injected in the mosquito and thus prevent the subsequent stages responsible for the disease and transmission of the infection to others. Humans and animals have been protected from malaria by immunization with irradiated sporozoites, but this method is impractical because of the limited supply and instability of sporozoites.

Recently, monoclonal antibodies have helped to define the structure of a parasite surface protein which appears to interact with host cells and which may therefore constitute a basis for protective immunity. The circumsporozoite surface protein was identified first in *Plasmodium berghei*, a parasite of mice (Yoshida, Nussenzweig et al., 1980). The gene for the circumsporozoite protein of *P. falciparum* has been cloned and sequenced (Dame et al., 1984) which soon should lead to mass-production of the specific antigen for immunization against sporozoites of this malaria parasite in humans. (It is of interest that children who carry a gene for sickle cell haemoglobin (trait) or for thalassaemia or are deficient in the red blood cell enzyme glucose-6-phosphate dehydrogenase (G6PD) survive *Plasmodium falciparum* infections better than 'normals' do; also West Africans who lack Duffy antigen or red blood cells are resistant to *P. vivax* infections (see Omenn and Motulsky, 1978).)

Viral hepatitis

The two main types of hepatitis virus, hepatitis A and hepatitis B, are endemic throughout the world but are especially prevalent in tropical and developing countries, where they are spread by poor environmental sanitation. Hepatitis A is transmitted by the faecal-oral route; the disease does not become chronic and no chronic carriers exist. The virus can be grown in cell culture; both inactivated and live attenuated hepatitis A vaccines are being tested.

Hepatitis B virus (HBV) is transmitted through inoculation of blood or blood products containing virus, or through close personal contact with a virus-positive person. It is a pathogen of major importance with an estimated 200 million carriers in the world. Acute illnesses vary in severity, and infection often persists, especially in children infected perinatally. Persistent hepatitis B may cause progressive liver disease (including chronic active hepatitis and cirrhosis), as well as primary liver cancer, one of the world's most common fatal tumours. The hepatitis B virus is an enveloped virus containing circular, double-stranded DNA. The virus cannot be grown in cell cultures but the entire genome has been sequenced and cloned in bacterial and eukaryotic cells. Inactivated hepatitis B surface antigen purified from the infectious plasma of asymptomatic human carriers has been used to produce vaccine to prevent both hepatitis and hepatitis-related hepatocellular carcinoma; field trials are taking place, for example, in Taiwan (Beasley et al., 1981). 'Second generation' vaccines prepared from HBV polypeptides are being tested for safety, immunogenicity, and protective efficacy in susceptible chimpanzees (Deinhardt and Gust, 1982), an approach designed to avoid risks associated with virus isolation from infected humans.

Priorities for immunization against hepatitis B may differ among geographical regions or countries, due to differing epidemiological patterns, cultural practices, socioeconomic and environmental factors. Studies to assess the prevention of liver cancer will require surveillance of subjects for many years. Longer term research is directed at developing vectors for DNA sequences coding for the immunizing HBV protein, and cloning of the HBV surface antigen in yeast.

Leprosy

The causative agent of this chronic affliction is an acid-fast bacillus (*Mycobacterium leprae*) which has not yet been cultivated *in vitro*. Leprosy presents a striking variety of forms, depending on the type of immune response of the individual. In a significant proportion of cases, there are serious sequelae, and a strong prejudice persists against the disease, leading to social ostracism of patients. Control of leprosy through mass treatment is limited by its long incubation period, its insidious onset, its chronic course, and the need for prolonged treatment. There has been tremendous progress recently in the immunology of leprosy — it is part of the WHO Expanded Programme on Immunizations — including development of a serologic test using a glycolipid specific for *Mycobacterium leprae* (Young and Buchanan, 1983). Large quantities of *M. leprae* have been grown in the nine-banded armadillo, stimulating a search for a vaccine. Purified preparations of *M.*

leprae produce good delayed-type hypersensitivity in mice and guinea pigs; M. leprae plus BCG can induce cell-mediated immunity in some patients and contacts. Field trials of vaccine preparations are being considered (Nordeen and Sansarrico, 1984).

Other vaccine preventable diseases

A study group on vaccines concluded that nine human diseases, selected because of the magnitude of the problems they cause and likelihood of significant payoffs within 5 to 20 years, deserve priority research consideration: malaria, bacterial enteric diseases (diarrhoea), respiratory diseases, tuberculosis, leishmaniasis, dengue fever, Japanese encephalitis, chlamydial infections and rabies (National Research Council, 1982).

Acute respiratory infections are major causes of morbidity and mortality worldwide, with over two million deaths per year. Infectious agents other than bacteria have been estimated to be responsible for 95% of all cases of acute disease of the upper respiratory tract and a considerable, though lesser, proportion of cases of disease of the lower respiratory tract. Several families of viruses are responsible: the common cold viruses; influenza A and influenza B; three types of parainfluenza viruses; respiratory syncytial virus; endemic and epidemic adenoviruses. Vaccines provide the logical and most obvious means of preventing viral respiratory diseases, but they have generally been unacceptable or disappointing, owing to adverse reactions or poor protective immunity. Research is needed on the antigenic composition of these viruses, the delineation of true immunogens from irrelevant viral antigens and toxic components, and the pathogenesis of the major viral respiratory diseases. For example, new epidemic and pandemic strains of influenza seem to arise from combinations of animal and human influenza viruses, a source of antigenically new strains for which there is no pre-existing immunity.

Rotavirus infantile diarrhoea may be responsible for much of the morbidity and mortality in very young infants in developing countries. Rotaviruses which infect humans are difficult to propagate in cell cultures; detection currently requires electron microscopy, though infections can be diagnosed by demonstrating rotavirus antigen in stool specimens or by measuring serological responses. Recently an oral live rotavirus vaccine gave promising results for immunogenicity and safety in a small number of seronegative adults and young children (Vesikari et al., 1983).

Work on attenuated live vaccines against dengue and dengue haemorrhagic fever is hampered by poor growth of the virus in cell cultures, lack of knowledge of its molecular biology, and the need to rely on primates to predict virulence in humans. An overriding concern in the development of dengue vaccines is the risk of potential immunopathology from the immunization itself (WHO, 1983c).

Current research on chlamydial diseases is focused on identification and characterization of immunogenic antigens of *Chlamydia trachomatis* and production of candidate protective antigens using biotechnology. Post-infection cell-mediated immunity has been shown in humans, but there is no vaccine so far. A monoclonal antibody diagnostic test for *C. trachomatis* is available commercially.

Vaccines may not be a feasible approach to certain important infections, especially trypanosomiasis. Molecular biologists are finding the trypanosome a fascinating subject, because of its remarkable genetic mechanisms for foiling the immune response of the host (Kolata, 1984). These organisms switch surface coat antigens during the one- to two-week period during which antibodies are formed, leading to bursts of proliferation during the infectious phase before invasion of the central nervous system to produce sleeping sickness. The surface has approximately 10 million molecules of variable surface glycoprotein, and there are hundreds or thousands of genes for these surface molecules; about 12-14 are expressed first, then after about 5 days these are turned off in a matter of a few hours and the infection continues with expression of one gene after another. Chagas trypanosomes seem to have analogous surface molecules and gene expression signals. Clearly, drug treatments will be essential for these diseases. It is interesting that trypanosome infections produce immune changes that resemble changes seen with the Acquired Immune Deficiency Syndrome (AIDS), which is becoming a major new epidemic, and that suramin, a drug used against sleeping sickness, has been reported to be effective *in vitro* against the putative agent of AIDS (retrovirus HTLV-III) (Mitsuya et al., 1984).

DRUGS

The scientific community is constantly searching for safer and more effective therapies to treat acute and chronic diseases, and for pharmacological, physiological and behavioural means to prevent specific diseases from occurring in the first place. However, the development of safe, effective and affordable drugs for treatment and prevention of major health problems in developing countries is a mostly neglected challenge, partly due to limited basic knowledge and largely due to lack of adequate economic return on investment for pharmaceutical firms.

Essential drugs

Together with vaccines, drugs are the most widely used health care technologies in the world. Many drugs that have been available for a long time in developed societies are now being exported. In addition, new agents must be developed for many viral and parasitic diseases, and means of overcoming and preventing drug-resistance are needed. The challenge is to decide which drugs are essential and to find the will and the resources to provide them when and where they are needed.

In 1977, WHO started a peaceful revolution in international public health by asking a group of experts which drugs were really necessary to take care of most health problems. Despite opposition from some physicians and the pharmaceutical industry, a model list of essential drugs, published first in 1977 and revised in 1979 and 1983, has survived the test of time (WHO, 1977, 1979c, 1983a). The current list includes about 240 drugs arranged in 27 categories, such as anaesthetics, anti-infectives, blood products, hormones, vitamins and minerals. Most of them were of proven efficacy, with well-known therapeutic properties. Most were no longer protected by patent rights and could be mass-produced at a reasonable cost for patients. An important point to remember, however, is that the WHO essential drug list is only a guide; each country has to adapt the list of its particular requirements to the level of resources it can commit to a pharmaceutical programme and to the structure and capabilities of its health care delivery system. The WHO Action Programme on Essential Drugs and Vaccines does provide guidance to developing countries in setting up their own national programme on essential drugs, following seven steps: (1) formulate a national drug policy; (2) decide upon a list of essential drugs; (3) procure drugs, utilizing bulk orders and UNICEF and WHO assistance; (4) establish logistics of supply (WHO's goal is to have the 20 most-needed drugs within an hour's travel from an efficient drug supply management system); (5) educate health professionals and the public on proper use of drugs; (6) assure quality control for efficacy and safety; and (7) train staff in policy formulation, selection, procurement, use, production, quality control and regulatory control of drugs. An additional guideline should be emphasized: to remain well informed about new drug developments and to update the list of essential drugs periodically.

The concept of essential drug programmes has taken root in many developing countries. For example, the New Drug Policy in Bangladesh includes elimination of harmful and useless medicines; increased domestic production of essential drugs; a public distribution system; bulk importation of pharmaceutical raw materials from different sources at competitive prices; use of generic rather than brand names; and encouragement of locally organized applied drug research (Ghulaam Mostafa, 1984). Since 1982, manufacture of essential drugs has increased, the country depends less on imports, and prices have fallen. A list of 150 essential drugs has been identified; 12 drugs have been selected for use by the village-level health workers for the most common diseases in the rural areas, and 45 drugs have been selected for district-level health care complexes. Another 100 drugs serve specialized or complicated cases.

In Kenya, a new drug supply system was introduced in two pilot districts in 1980 and is intended to cover all districts by 1985, including the 80% of the population who live in rural areas (Steenstrup, 1984). A total of 39 essential drug items, selected to reflect the morbidity and mortality patterns in rural areas, are provided to the rural health centres and dispensaries in pre-packed ration kits. They are procured in generic form on the basis of their therapeutic effects and comparative cost. A massive training programme has been introduced for rural health workers. The new system has succeeded in attracting many patients away from district hospitals to rural health centres and dispensaries, at lower cost per patient.

Nevertheless, much more needs to be done. In Latin America, where as much as 40% of national health care budgets is spent on drugs, most of the population does not have access to drugs for the most common diseases. The number of pharmaceutical items on the market may be 80,000 in some countries; the great majority are duplicates or combinations of drugs with doubtful effectiveness or safety, or do not represent the agents of choice for major endemic diseases (Nightingale, 1984).

OTHER MEASURES

Oral Rehydration Therapy (ORT)

Acute diarrhoea affects nearly 500 million children annually and some 5 million die from dehydration and loss of electrolytes. Intravenous treatment of diarrhoeal diseases, mainly cholera, with saline solution was first used in 1832, but mortality rates remained very high until a scientifically balanced electrolyte intravenous solution was introduced in 1946. Since the 1960s, worldwide practical experience has demonstrated that early use of oral rehydration therapy can avoid the need for intravenous treatment, making ORT 'potentially the most important medical advance of this century' (*Lancet*, 1978).

Intestinal absorption of water and salts is impaired during an attack of diarrhoea, but absorption of glucose and some other substances remains normal. Since the absorption of glucose greatly enhances the simultaneous absorption of water and electrolytes, a solution of glucose, salt and other ingredients can overcome the absorption problem. The standard WHO/UNICEF formula per litre of water now is: sodium chloride 3.5 g, sodium bicarbonate 2.5 g, potassium chloride 1.5 g and glucose 20.0 g. The solution is given alone for the first six or eight hours of treatment so as to achieve rehydration, then continued, together with breastmilk and appropriate fluids and foods, for the next few days to maintain correct hydration and improve malnutrition. At least 40 countries are formulating well-defined oral rehydration programmes, and a number already have encouraging results, as in Egypt, Indonesia, Thailand, Honduras, Philippines and Tunisia (Tuli, 1984). However, many questions must still be answered if there is to be control over the diarrhoeas. Laboratory work is needed to determine virulence factors and develop vaccines or drugs to inactivate them; field epidemiological and clinical studies are needed to define better the agents of importance in various areas of the world, among specific age-groups, and during specific seasons of the year.

Contraceptive technology

Advances in contraceptive technology over the past 20 years include development and refinement of oral contraceptives and intra-uterine devices; development of an injectable hormonal contraceptive; legitimation and increasing use of the condom as an acceptable public programme method; simpler methods of female sterilization; mass use of vasectomy in several male-dominated societies; and greater knowledge of the sequence of events during the menstrual cycle and consequent refinements in the rhythm method. In addition, several new contraceptive methods are in the research and development stage, such as anti-fertility vaccines, male contraceptives, and reversible sterilization procedures. The technology for safely performing abortions has also improved. The development of national family planning programmes is probably one of the major social revolutions of the twentieth century (National Research Council, 1979).

Improved contraceptive technology is widely recognized as a high priority need in both developed and developing nations, although there are conflicts over differing benefit risk considerations. For example, the two currently available injectable contraceptives give rise to disturbance of the menstrual cycle and thus are unacceptable to many women. Clearly, better compounds are needed, but pharmaceutical industry investments are limited by the lack of demand for such products in developed countries. As a result, the WHO Special Programme on Research in Human Reproduction and Population Control, under the direction of the WHO Advisory Committee on Medical Research, has challenged scientists in academic institutions with this task. About 300 new compounds have been prepared in university laboratories in eleven different countries; these compounds have been tested in animals, and two have been selected for study in humans. In a departure from the past WHO policy in patents, the Special Programme has gained five patents on injectable contraceptives, and several are pending (Standley and Kessler, 1983).

Health care delivery

HEALTH CARE PROVIDERS

Neither high technology biomedical initiatives nor simple, low-cost applied technologies for health can be implemented in developing countries without appropriately trained, well-organized people — both professionals and paraprofessionals.

Professionals

The training, deployment and utilization of professional health personnel in developing countries tends to be patterned after that of the industrialized countries. Education programmes for both physicians and nurses have been imported from Europe and North America, and large numbers of physicians and nurses have been sent to Europe or North America for training that is often inappropriate for the community settings and teamwork requirement to which they return. Selection of students for health professional training has tended to favour the elite, who usually then choose to practise in familiar urban settings and in medical specialities relevant to their educational experience (Institute of Medicine, 1980). The supply of health professionals in most developed countries is very limited, largely because of limited finance and the tendency to direct funds to curative care, mainly in urban hospital facilities (World Bank, 1975a). The World Health Organization reported that in 1970 the number of physicians per 10,000 population was 8 for the world as a whole, 15 for Europe, 16 for North America, 1 for Africa, 3 for Asia and 6 for Latin America (WHO, 1975). Even to maintain these 1970 physician population ratios to the year 2000 would require a 115% increase in the total number of physicians in Asia and a 136% increase in Latin America and Africa.

In all countries, interest in the relationship between educational programmes for the health professions and the actual practice of health care increased during the 1970s. In the developing world, concern over scarce resources led to questioning of the traditional approaches to medical education, while in the industrialized world there has been concern over the rapidly increasing cost of health care. One result was the formation of a Network of Community-Oriented Educational Institutions for Health Sciences (with its headquarters at the Rijksuniversiteit Limburg in Maastricht, in the Netherlands), which now includes approximately 40 medical schools, half in developing countries and half in developed countries. These medical schools place special emphasis on population-based primary health care in their education, research and health care activities (Greep and Schmidt, 1984). For example, at the University of Ilorin, in mid-western Nigeria, a regular exodus of students and staff into the rural areas takes place twice a year. Small groups of students, each accompanied by two staff members, settle for a month in a village, where they study disease patterns, make epidemiological surveys, and contribute to local health care. During their six-year period of study, students are exposed to community experiences for the equivalent of one year and are sensitized to the primary health needs of the community, appreciating diseases in their social, economic and cultural contexts.

Paraprofessionals

The wide discrepancies between needs and numbers of professional health personnel have stimulated training and use of auxiliary health workers, paraprofessionals, multipurpose workers, and assistants in nursing (including traditional midwifery), environmental sanitation, pharmacy and curative medicine. In many instances, these health auxiliaries have assumed responsibilities traditionally assigned to health professionals. Paraprofessionals are often undersupervised, overworked and inadequately trained. For example, instruction may not even include simple emergency surgical procedures, yet they may have to deal with complicated obstetrical cases, fractures and other injuries.

Community health workers are a link to the formal health care system. Such workers are volunteers or part-time workers from the local community paid what the community can afford. In well-run programmes, community health workers are given basic instruction in general health concepts, emergencies and referrals, common diseases, health education, nutrition, family planning techniques, environmental sanitation and basic recordkeeping. These skills, combined with their community rapport, can make such personnel very valuable in providing health services for all by the year 2000. A recent review of programmes in several developing countries demonstrated that there are major problems in selection, recruitment, training, functions, remuneration and career prospects, but the most serious problems involve inadequate support services (Ofosu-Amaah, 1983). In general, national planning of community health worker programmes is lacking, and professional health workers appreciate neither the roles and functions of the community health workers nor what can be done to support them.

Indigenous healers

Over the centuries, societies and communities have evolved countless methods of traditional healing to cope with physical and psychological needs and to accommodate local social structures. Most traditional healers provide a combination of preventive and therapeutic health care in villages and homes with a combination of traditional and modern medications.

In most developing countries, governments are attempting to replace these traditional healers with scientifically trained allied health workers; the numbers and distribution of trained allied health workers are far below the needs, however, and many rural people have greater confidence in, are located closer to, and continue to rely on traditional healers. Recognizing the realities of this situation, some developing countries, in cooperation with the World Health Organization, have initiated programmes that attempt to integrate modern and traditional medicine by training traditional healers in scientific principles and techniques. The major barrier to utilization of traditional practitioners and beliefs is the unwillingness of physicians and other health professionals to allow traditional drugs and practices to be made part of, or to exist alongside, the modern medical system. At best, medical professionals try to integrate limited aspects of such systems into their own organizations, but usually they fail to recognize the context within which traditional systems operate and thus inadvertently render 'borrowed' components ineffective.

PRIMARY HEALTH CARE

The temptation has been strong for developing countries to try to adopt the high-technology, urban hospital-based model of health care provision of the industrialized countries, even though their problems are very different, and even though that model is beginning to be called into question even in the West. The alternative is the less costly approach of primary health care projects, using appropriately trained professionals and nurses to deal with basic health problems, often in rural areas, leaving doctors and hospitals to concentrate on the more complicated cases that require specialized treatment.

A model primary health care system has been described in Dhende Mao, India, where a nurse with two years of government-sponsored training provides prenatal care, nutritional advice, vitamin supplements, immunizations and antibiotics. The nurse is the middle level of a three-tier 'health pyramid'; she receives patients who cannot be helped by a first-aid nurse, usually a man who runs a fever treatment centre in his home. If the clinic nurse needs further assistance, she refers the patient to a doctor some 20 km away. These referrals facilitate treatment of the more difficult cases and enhance the credibility of the three-tier system.

Examples of success rates and costs of primary health care in projects in four different countries are summarized in Table 10. In Narangwal, India, three projects were compared, providing either health care, nutrition supplements or both, using illiterate health workers. Similar villages nearby, lacking significant primary health care services, served as controls. The cheapest and most effective approach in reducing child death and disease (at an estimated US $ 60-75 per life saved) was the health-care-only project. At

Miraj, India, with a per capita cost of only US $0.85 per year, infant mortality was reduced in three years from 6.8% to 2.3%; the number of children immunized against the basic childhood diseases increased from about 5% to 85%; 97% of mothers received prenatal care, and the birth rate was reduced. At the Albert Schweitzer Memorial Hospital in Haiti, for less than $3.85 per person per year, a programme that used community health workers cut rates of mortality to one-sixth of the national levels, and 85% of the children were immunized for diphtheria, tetanus and whooping cough, compared with only 15% nationally. These primary health care demonstration projects are encouraging because local communities were involved in their operation, the projects became substantially self-supporting, and curative and preventive care were combined.

Table 10. Primary health care projects: success rates and cost.

Project	Infant mortality rate		Annual cost
	Project area (%)	Control area (%)	per capita (1983 U.S dollars)
Guatemala, rural (1970-2)	5.5	8.5	8.90
Haïti, rural (1968-72)	3.4	15.0	3.85
India, Jamkhed (1971-6)	3.9	9.0	1.55
India, Narangwal (1970-3)			
medical care	7.0	12.8	2.40
nutritional intervention	9.7	12.8	4.80
combined	8.1	12.8	3.60
Nigeria, Imesi (1966-7)	4.8	9.1	3.60

Source: Chandler, 1984.

A major achievement of primary health care projects has been a large increase in families practising family planning. In Jamkhed, India, family planning participation rates increased from 10% in the surrounding area to 50% in the project. A couple's willingness to practise birth control strongly depended upon their past history of infant and child losses — as health care improved the chance of having children who survived, the number of parents using contraceptives increased quickly.

Reports from other prototype primary health care projects are discouraging, however. A survey by the United States Agency for International Development of 52 primary health care projects that it funded, noted such common problems as poor administration, lack of minimally educated personnel, lack of medicines and supplies, poor communication, poor transportation and inadequate follow-up training (Burns Parlato, 1982). While the knowledge and technology to overcome difficulties in the delivery of primary health care services need to be improved, politicians, planners and decision-makers must be prepared to put available methods into action.

COMMUNITY MOBILIZATION

The term 'community' usually refers to a group of people living in the same geographic area, such as a rural village or an urban neighbourhood, but it can also mean a religious, ethnic or occupational group whose members interact but do not all live in the same geographic area. Community mobilization efforts are gaining ground in concert with increases in schooling and literacy, agricultural productivity, standards of living, transportation and communication. In the context of health care, community mobilization may include gathering and processing local medicinal herbs, assisting women in childbirth, learning how to use latrines, and maintaining newly provided village water supplies. Community mobilization can have the following benefits: support from the local power structure; more accurate assessment of needs; decreased dependency on limited external resources; better use of the indigenous health care system; recruitment of appropriate community members for the position of community health care workers; increased service coverage; increased service utilization; active participation of community members in preventive and related health-promoting activities. Community participation in primary health can also help a village achieve self-reliance and social awareness.

Perhaps the largest obstacle to overcome in mobilizing community support for primary health care delivery is the lack of governmental commitment. Many developing governments have not given high priority to health. When they have asked for assistance, the requests often have been for advanced hospital technology (like heart-lung machines, or lasers, or tomographic scanners), rather than for assistance aimed at the health of low-income groups. Developing country leaders generally have requested the kind of assistance that donor agencies have been prepared to provide and that elite minorities seek for themselves. Both developing countries and advanced countries have been influenced by the Western emphasis on hospital-based, high-technology medicine (Bell, 1980). Only in recent years has it become plain that another approach based on primary health care is essential for the poor majority in both rural and urban communities.

ROLE OF WOMEN

The roles, education and status of women interact in a triangle of health, population and development. Women outlive men in developed countries, but in many developing countries women have a higher mortality rate than men, especially in South Asia and North Africa; high female mortality occurs at two peak periods, in early childhood and during the reproductive period. Investigations confirm the common assumptions that boys are fed and treated better than girls. Women's access to health and family planning services depends to a large extent on factors that affect the population as a whole — such as the availability of transportation, trained health workers, supplies and equipment — but despite its importance it often rates a low priority within national health programmes.

In many developing countries there are strong pressures against fertility control: a woman's status depends on her childbearing performance; she can be a social outcast if she has no children, and barely acceptable if she has only one or two, bringing her husband's virility into question. Yet motherhood often conflicts with the need to work, whether in subsistence agriculture or, despite poor pay, for cash to support their families and themselves. Despite the importance of childcare, breastfeeding and weaning patterns on a child's health and nutritional status, women may be impeded by development-related barriers such as agricultural shifts to cash-cropping, the entrance of young mothers into the wage labour force, migration and separation, and land settlement schemes in which women are isolated from their kin and neighbours. Furthermore, training in agriculture and industrial development and transfer of technology have been geared towards men, despite women's extensive role in production (this is normally neglected in statistics, yet in Africa, for instance, 60-70% of agricultural work is performed by women). This has meant an easing of men's burdens and an increase in productivity for their part in the workload, while women have been left with even greater manual tasks and lower productivity, in addition to family responsibilities.

It is important that the entrance of women into the labour force occur under conditions that permit them to limit childbearing and improve their own health, in turn improving the health status of their families and communities.

Female literacy is an important aspect of women's role in health care, probably because literate mothers can understand and practise basic elements of disease prevention and follow treatment plans more completely for their infants and children. In a case study of health status in Kenya, Mosley (1983) has shown that maternal education and level of poverty explain almost all the variance among infant mortality rates. In a study of 41 African countries, the strongest associations to health status were the variables of literacy and per capita GNP. Among a number of variables examined for effects on child survival in Nigeria, maternal education had a strong effect, and father's occupation was significant, while urban versus rural residence was negligible (Caldwell, 1979). The relationship between infant mortality and female literacy is shown in Table 11 for selected countries.

Table 11. Infant mortality and female literacy in selected countries.

	Infant mortality, 1981 (%)	Female literacy, 1980 (%)
Burkina Faso	21	5
Afghanistan	20	6
Ethiopia	15	5
Bolivia	13	58
Nigeria	13	23
India	12	29
Pakistan	12	18
Saudi Arabia	11	12
Tanzania	10	23
Honduras	9	62
Brazil	8	73
Mexico	5	80
Philippines	5	88
Thailand	5	83
China	4.1	66
Yugoslavia	3.1	81
Costa Rica	2.7	92
U.S.S.R.	2.6	98
U.S.A.	1.2	99
Japan	0.7	99

Source: Chandler, 1984.

Research and development in health care technology

'HIGH-TECHNOLOGY' VERSUS 'HALF-WAY TECHNOLOGY'

Lewis Thomas, Chancellor of the Memorial Sloan-Kettering Cancer Center in New York, has described technology as the driving force of modern medicine and argued that human health can be improved at an affordable cost only by advancing technology through basic research. He distinguishes high technology from half-way technology (Thomas, 1983, 1984). 'High technology' includes vaccines and drugs that prevent or cure specific diseases: they work through detailed understanding, or at least empirical evidence, of underlying mechanisms of cause and effect. Polio vaccine, penicillin, and the eradication of smallpox are great triumphs of high technology. The research involved may take many years and the costs are not trivial. In the long term, however, the applications may be extremely efficient and inexpensive: polio has been eradicated and measles cases have fallen by 80% in the United States of America as a result of vaccines that cost a few cents each. Similar breakthroughs could be achieved by current research into malaria and cholera vaccines. More than laboratory research is required, of course — vaccine development is a complex and lengthy process, involving field tests and logistics for wide utilization.

'Half-way technology' refers to those systems or devices that relieve symptoms, but do not cure disease. The iron lung for polio victims, coronary bypass surgery, and many cancer treatments fall under this heading. In contrast to vaccines and antibiotics, it can be both expensive and inefficient, because equipment, tests and treatments are often used without adequate consideration of their appropriateness and cost. Nevertheless diagnostic, immunologic, pharmacologic and hospital techniques in this category can provide vast benefits if they bring improvements to hospital technologies, which account for half of the health budgets in developing countries and one-third in industrialized countries (Chandler, 1984).

Given the inequitable distribution of wealth and resources in the world, it is not surprising that the World Health Organization estimates that 97% of the world's biomedical research funds are spent on diseases primarily of interest to the developed countries and only about 3% on the diseases primarily of importance in developing countries (Institute of Medicine, 1980). Furthermore, discoveries which could be of vital importance to developing countries may be neglected because they are not relevant to Western problems. The principle on which ORT is based was discovered in the United States of America, but ignored because no use was seen for it, and it had to be rediscovered by researchers working on diarrhoeal dehydration in Bangladesh. Biomedical research ought not to be duplicated all over the world while the most basic tools (stethoscopes, scales, antibiotics, refrigerators, X-ray machines) are absent in many areas of developing countries — hence the need for systematic technology assessment to determine the allocation of scarce resources, a problem faced by all countries (Banta, 1984). As documented in numerous cases in this chapter, technology offers major hope for improving the health of people throughout the world, yet even in the United States of America, where almost 11% of the GNP is spent on health care services, many needs are unmet.

Policies are needed to address the four stages of health care technology development: (1) research and development, including adaptation to new settings; (2) evaluation of various technologies and their fit in health care systems; (3) regulation of safety and efficacy; and (4) decisions on utilization of investment resources. Adoption of new equipment or procedures for screening, diagnosis and treatment usually proceeds incrementally without any process for gathering objective data or comparing cost-effectiveness for meeting health status goals. The Pan American Health Organization (PAHO) and the World Health Organization, as well as many national organizations, are trying to stimulate systematic programmes to identify and monitor emerging technologies and new applications of existing technologies. Such a system provides inputs for analyses and then dissemination of conclusions to all interested parties.

IN-COUNTRY RESEARCH

One priority for health care technology research within developing countries is the generation of new knowledge of endemic diseases. The extent of existing knowledge applicable to control of the diseases that beset the developing world's populations varies greatly. For most, there are needs for more effective therapy and for simple, effective methods for prevention or for delivery of medical care, as outlined above. Most health problems of developing countries will require three concurrent or sequential lines of investigation: fundamental research, adaptive research and field testing.

Fundamental research consists of studies designed to provide new insights into the natural history, epidemiology, clinical manifestations and pathogenesis of diseases in diverse settings. Studies of genetically, socially, culturally and behaviourally conditioned diseases and responses to therapy are also included in fundamental research. Examples include:

— research on the biology of the etiologic agents and the pathogenesis of certain diseases, such as South American trypanosomiasis (Chagas' disease) and onchocerciasis, with a view to developing more effective drugs and vaccines

— studies of immunologic response to parasitic infections to provide insights into immediate and delayed hypersensitivity reactions that occur in these and other diseases, to discover means of blocking such responses and, thereby, the disease symptoms

— long-term epidemiological studies of the relationships between schistosomal infection and disease to identify the relatively small proportion of infected hosts who are at risk of developing overt disease

— research on the clinical manifestations and outcome of common mental illnesses and of culture-bound syndromes, for better understanding of relationships between culture and mental illness

— pharmacologic investigation of traditional drugs and remedies used by local healers, to enhance their rational application.

Adaptive research is the transformation and application of established technology to suit local conditions, or modification of existing treatment modalities and public health practices to improve efficiency, increase patient compliance or reduce toxicity. Examples include:

— development and mass-production of inexpensive methods for oral rehydration therapy that can be used safely and effectively by mothers and semi-skilled local health workers

— methods for safe storage, supply and distribution of drugs and vaccines in the hot and humid conditions prevailing in most developing countries

— developing and testing of new dosage forms or methods of administration of effective drugs to make them safer and easier to use, and to ensure better compliance.

Field testing includes research designed to assess the effects of new methods of control or treatment in specific populations in which the responses may be influenced by genetic, cultural or environmental factors, and studies to evaluate modifications of existing technology. Examples include:

— assessment of the effect of new control programmes by comparing indices of health status in an area in which a new programme has been introduced with those in other areas in which it has not been introduced. Just as important is assessment of the current value of programmes in use throughout a country, but which may no longer be needed or are no longer effective

— comparison of costs and effectiveness of two methods of managing oral rehydration therapy: mass-distribution of packets and instructions to all mothers of infants prior to occurrence of diarrhoea, or selective distribution of packets by local health workers after episodes of diarrhoea have been reported

— comparative studies of outcomes, costs and potential toxicities of modern medicine and traditional folk practices to provide a rational basis for decisions on the integration of indigenous medicine into the primary health care delivery system.

There are many examples of the striking impact of the results of such in-country health care research (Institute of Medicine, 1980). Controlled clinical trials in Madras by the Tuberculosis and Respiratory Disease Research Unit of the British Medical Research Council and their Indian counterparts found that tuberculosis patients could be treated as successfully and much less expensively on an ambulatory basis as in the hospital, with no greater risk to their families. In a setting where both funds and hospital beds are extremely limited, these findings were of immediate benefit.

The multinational group at the Cholera Research Laboratory in Dacca, Bangladesh, developed oral rehydration procedures. WHO is now disseminating this life-saving technique worldwide through the Diarrhoeal Disease Control Programme. In India — with a small number of psychiatrists, a vast population, and severely limited resources for mental health — researchers at the National Institute of Mental Health in Bangalore is determining ways to make more effective use of the skills of traditional healers, who are found in every village; they are being trained to discriminate between the conditions for which their arts can be useful and those which require the tools of Western medicine, with promising preliminary results. Studies of several Chinese communities have shown that cultural influences on illness behaviour and doctor-patient transactions affect patterns of utilization, compliance and satisfaction, as well as therapeutic outcomes. Such studies can be useful to health care providers concerned with adapting Western medical practice to traditional cultures.

INTERNATIONAL COLLABORATION

The growth of scientific and technical capabilities in developing countries over the past three decades offers new opportunities to create cooperative international relationships between developing and more developed countries. Multilateral and bilateral assistance must play a significant role in helping developing countries to strengthen their own institutions. In the search for research initiatives for health technology collaboration, it is important for the developed countries to keep in mind these factors (National Research Council, 1979):

— needs and priorities of developing countries
— special concerns of developing countries, including the desire for self-reliance and greater autonomy in the management of their affairs
— potential mechanisms for transferring and adapting foreign technology to conditions found in developing countries
— ways to help developing nations draw upon worldwide science and technology in devising answers to their own problems, and contribute experience and advances for use by other countries.

Each developing country has unique scientific, technological and developmental needs. Moreover, developing countries are showing a growing interest in interregional cooperation. The developed countries must be prepared to offer and participate in programmes that aim at a wide range of needs, appeal to different countries and regions, and tap the varied interests and capabilities of their own people and institutions.

Health care technology is transferred from the developed countries to developing nations by many routes, including universities, visiting experts, books, journals, technical reports and health care products. Commercial transfer of health technology often involves proprietary knowledge, available only on terms negotiated with the owners of the technology being sought. In the private enterprise system of the industrialized nations, the ability to protect private skill and information is an important incentive for economic performance and technical advancement. Most proprietary technology can be purchased, and developing countries are increasingly adept at negotiating better terms for it. Much health technology is non-proprietary, and technology in expired patents that have never been brought to large-scale commercialization may, because of changing circumstances or particular needs in developing countries, be newly attractive.

Assuming that universities, government laboratories and industries are prepared to cooperate in transferring health technology and in seeking its adaptation to conditions in developing countries, the developing countries need the capability to assess, select, develop, adapt and apply the most useful technologies. Research and development collaboration in health technology can occur in various forms, as described below for some major organizations.

The World Health Organization has identified a 'revolution in research' as an 'irresistible logical process and a passionate protest against the doctrine of inequity and want' (WHO, 1980c). In recent years, WHO has emphasized research and research training, and the creation of collaborating centres and other opportunities for full participation of research scientists, in view of the fact that scientific discoveries and applications historically have depended upon the efforts of many scientists working in centres scattered all over the globe and exchanging the results of their scientific work.

In 1958, WHO convened the Advisory Committee on Medical Research (ACMR). During the first decade of the ACMR, modest funds were made available to coordinate research in existing laboratories, involving some 740 investigators in 430 institutions in 77 countries (Kaplan, 1983). A network of international and regional WHO reference centres started work in the fields of viral, bacterial and parasitic diseases, environmental health, mental health, radiation health, nutrition, vector biology and control, and viral and health statistics. By 1968, it had become evident that new approaches were needed to meet serious mortality and morbidity problems in developing countries. Since it is impossible to control all diseases at once, efforts were targeted. The Special Programme for Research and Training in Tropical Diseases (TDR) was established with co-sponsorship by the United Nations Development Programme (UNDP) and the World Bank. A consortium of donor agencies helped to raise extra-budgetary funds — an approach which had earlier generated funding for the Special Programme on Research, Development and Research Training in Human Reproduction. Task forces of leading scientists in each of these fields were charged with determining research approaches, identifying collaborating laboratories, and helping with the creation of new units. The TDR programme has been one of WHO's most successful research undertakings.

The problem of health services research has plagued the ACMR almost since inception. Everyone seems to recognize its importance, but no common methodology has emerged and each country differs in its needs, resources, cultural habits and other factors, especially local political factors. The encouragement of research in mental health has evolved from a few reviews of mental health research to an expanded programme of research and training in bio-behavioural sciences and mental health, with emphasis on primary health care, the psychosocial aspects of rapid social and technical change, and health-promoting behaviour (Kaplan, 1983). The ACMR also did valuable work in guiding WHO's policy on such matters as recombinant DNA research and the safety requirements for laboratories handling dangerous micro-organisms.

The Bureau of Health and Population of the United States Agency for International Development (AID) is presently supporting applied and basic research in tropical diseases, primary health care delivery, and water and sanitation, including the following activities: the immunology and biochemistry of malaria (*P. falciparum*); development and testing of anti-sporozoite and possibly anti-merozoite vaccine products; oral rehydration therapy; a seven-nation effort in West Africa designed to control the black fly vectors of onchocerciasis; contributions to the Tropical Disease Research programme of WHO; primary health care projects; and development, evaluation and dissemination of contraceptives, including a subdermal contraceptive implant, a new microsphere injectable contraceptive, advanced IUDs, and non-surgical sterilization. The Walter Reed Army Institute of Research, the National Institutes of Health (NIH) and the Centers for Disease Control (CDC) are other American agencies involved in research in the field of tropical diseases.

The United States National Research Council's Board on Science and Technology for International Development (BOSTID) offers grants to developing country institutions for research in selected areas of science and technology that are important for economic development. The programme, which began in 1981, supports research in three health areas (mosquito vector field studies, rapid epidemiologic assessment, and acute respiratory infections in children) as well as three agricultural areas (Bell, 1984). For each area, an organizational meeting of expert scientists from the developing countries, international organizations and American institutions sets detailed objectives and research priorities and recommends the type of international collaboration that seems most useful. The recipient institution contributes partial support of the research staff, laboratory and or field testing facilities, vehicles and administrative services.

The Rockefeller Foundation has developed a three-part programme. Under the 'Great Neglected Diseases of Mankind', some outstanding scientific groups around the world have been stimulated on development of vaccines, drugs, diagnostic tests, and appropriate targeting of therapy for malaria, schistosomiasis, diarrhoeal and other diseases, in collaboration with 22 developing countries. The 'Health of Populations' component seeks to foster the discipline of clinical epidemiology in medical schools, and in resource and training centres in the developing world, with an initial emphasis on Asia. 'Coping with Biomedical and Health Information' facilitates dissemination of biomedical and epidemiological information, through a regional library system in Latin America and through the key medical schools and their libraries in China.

The Walter and Eliza Hall Institute of Medical Research in Melbourne, Australia, an independent charitable institution largely funded by the Australian National research Council, has conducted important research in parasitology, immunology and cell biology. Researchers collaborating with the Institute of Medical Research of Papua-New Guinea have overcome one of the major obstacles to producing a malaria vaccine through their work on surface antigens of *P.falciparum*. Thanks to agreements negotiated with Australian commercial interests, Papua-New Guinea will benefit from these vaccines at low cost.

RESEARCH PERSONNEL

New knowledge from research is more likely to be incorporated into a country's health care delivery system when its own citizens have participated in generating that knowledge — consequently, development of expertise at both the professional and technical support levels should be an important objective of health research in developing countries. Unfortunately, however, social attitudes and economic conditions in many developing countries do not encourage talented young people to elect careers in research: research does not usually confer prestige, and developing countries are frequently unable to provide job security or appropriate facilities and equipment for research. As in developed countries, the greater financial rewards and higher social status from the private practice of medicine or positions in industry attract young physicians and biomedical scientists.

As emphasized in the Institute of Medicine report (1980), training should be relevant to the health problems of the developing country and not primarily tied to new and expensive technology that may be inappropriate or unavailable. The training of epidemiologists, for instance, is particularly pertinent to the needs of developing countries, where the magnitude of the most serious health problems has often been only crudely estimated; descriptive epidemiology is particularly important for determining health programme priorities. Medical research can strengthen the skills of host country staff in laboratory work, recordkeeping and data processing. Health workers who become experienced in clinical trials are invaluable for field investigations, obtaining patient cooperation, keeping patients under observation, and noting unexpected reactions to therapy. Participation of these individuals in the health care delivery system can help to improve the quality of care and promote application of research findings to practice in both urban and rural areas.

The training of scientific talent *within* the developing countries, of which there are already successful examples, deserves high priority. Graduates of a three-year Master of Medicine degree course at Makerere University in

Uganda, established with an international faculty, in 1966 at the request of three East African countries, now hold responsible positions in research institutions throughout Africa. The research programme at the Institute of Nutrition of Central America and Panama (INCAP), which is supported by the Pan American Health Organization, illustrates the capacity of a regional organization to train young scientists from member nations. INCAP's regional laboratory courses have been successful in bringing modern microbiological, immunological and recombinant DNA techniques to scientists in developing countries. Such regional arrangements are less vulnerable to political changes within a single country, and can draw from a larger pool of eligible candidates.

One of the problems of extended out-of-country training, either in a regional institute or in a developed country, is that there is too often no provision for those who have been trained elsewhere to continue scientific careers on returning to their own country. Prior arrangements for research careers at the conclusion of training would encourage young medical scientists in developing countries to elect careers in research and to return after their training.

EXAMPLES OF APPLIED RESEARCH

Low-cost, effective health care technologies for use in primary health care represent a special need in applied research. Although governmental agencies have been involved in the development of such technologies, a major impetus has come from private, non-profit voluntary groups. Examples of such groups are PIACT (Program for the Introduction and Adaptation of Contraceptive Technology) and PATH (Program for Appropriate Technology in Health), based in Seattle, Washington. PIACT has affiliates in eight countries (Indonesia, Bangladesh, Mexico, Nepal, the Philippines, Sri Lanka, Thailand and Canada) and works with local organizations in several other countries to identify and strengthen indigenous resources for production, procurement, consumer participation and distribution of information about contraceptives and their uses, especially in formats suitable for illiterate or semiliterate populations. PATH seeks to bridge the three independent variables that affect the quality of primary health care programmes: the product or technology, the system through which it is made available, and the consumer or recipient for whom the service is designed. The following are three examples :

In order to prevent the use of degraded measles vaccine, and consequently ineffective immunizations, PATH has adapted a time- temperature indicator paper. This product contains a highly sophisticated chemical with well-characterized kinetics of polymerization, licensed from the Allied Corporation. With exposure to heat, the polymerization reaction results in a nonreversible change of colour from pink to red to black. This indicator is attached to each vial of vaccine; when critical cumulative exposure of time and temperature has been reached, its black colour reveals that the vaccine is no longer suitable for use. Even if a more stable measles vaccine becomes available, the time-temperature marker technology has many other applications in quality control in assuring that the 'cold-chain' is intact for biological products and foods that require refrigeration.

To be fully effective, oral rehydration therapy requires development of alternate dosage forms to minimize error and enhance acceptance; improvement of packaging; and development of explanatory materials for health workers and mothers. Packets of oral rehydration salts typically supply ingredients in powder form for a litre of solution (see p. 00). Since litre containers may not be available in some rural areas and because of cultural tendencies not to recognize loose powder as medicine, PATH has developed a tablet to dissolve in 150 cc of water (5 oz cup). It is thought that the tablet will be more readily perceived as medicine and that the smaller dosage will be used promptly, thus reducing the risk of bacterial contamination of any unused solution. The tablets are currently being field tested to determine their shelf life under various packaging and developing country conditions.

A major constraint to the wide use of an existing filtration method for diagnosing urinary schistosomiasis in the field has been the high cost of the filter. The cost per test has been reduced from US $0.60 to less than $0.05 with a transparent, easy-to-read millipore filter kit.

Conclusions

HEALTH GOALS

Improved health of people throughout the society must be recognized both as a measurable goal of economic development and as a requirement for achieving economic development. More resources are justified, and better management is needed. Dominant problems are communicable diseases, malnutrition, and too-rapid population growth.

Improvements depend upon a well-balanced combination of public health systems and personal health care systems. In general, control and prevention of diseases, when feasible, is much more cost-effective than treatment; this approach depends on national leadership and local cooperation. Development and maintenance of adequate, safe water supplies and effective sanitation facilities are crucial for control of the major vectors and agents of disease, and for healthy daily living. Malnutrition and diseases worsen the adverse effects of each other, so that improvements in the quantity and quality of food for all groups in the population, and especially for women and children, are critical to health goals. A comprehensive effort combining immunizations, breastfeeding, oral rehydration therapy, periodic surveillance of growth, family planning, better nutrition and safer water could bring mortality and morbidity rates for infants and children down toward those of industrialized countries.

With economic progress, the burden of diseases and disability shifts toward the pattern associated with industrialized countries, owing to adverse personal behaviours (especially cigarette smoking) and social behaviours reflected in urbanization, overcrowding and pollution. Developing countries have an opportunity to avoid some of the misfortune of countries which entered the industrial period decades or generations earlier.

MEASURES FOR IMPROVING HEALTH

Fundamental research, applied research and field tests of vaccines, drugs, test kits and other technologies for public health and health care are desperately needed. These efforts require much collaboration between appropriate personnel in developing and developed countries, well focused on the problems and conditions in the developing countries. Resource constraints and cultural barriers must be taken into account, so that R&D and field strategies can be developed and efforts can be targeted to accomplish as much as possible with the means available.

The WHO Expanded Programme on Immunization (against measles, pertussis, tetanus, polio, diphtheria and tuberculosis), the WHO Special Programme for Research and Training in Tropical Diseases (malaria, schistosomiasis, trypanosomiasis, filariasis, leishmaniasis and leprosy), and the WHO Special Programme on Human Reproduction and Population Control all represent well-coordinated, well-targeted, scientifically well-grounded collaborative efforts deserving of support and expansion.

Advances in the modern fields of genetics and immunology permit scientific and technological developments worldwide that may help developing countries to make rapid progress against age-old diseases, including diseases such as malaria whose vectors and parasites are becoming resistant to current agents. In particular, technologies of genetic engineering using recombinant DNA and cell fusion are being applied very successfully to the production of vaccines on an industrial scale. A range of products have either already been produced or will shortly be available using these methods: vaccines (hepatitis B, herpes simplex, cholera), hormones (insulin, growth hormones, beta-endorphin), enzymes and proteins. The United Nations Industrial Development Organization (UNIDO) has recently launched a programme on industrial production of biologicals to provide technical assistance to developing cuntries wishing to start domestic production of vaccines. Algeria, Brazil, Colombia, Cuba, Egypt, India, Indonesia, Mexico and Pakistan have already inaugurated extensive programmes of domestic vaccine production.

TRAINING, RESEARCH AND SERVICE DELIVERY SYSTEMS

Capabilities within the developing countries to train allied health workers, professional health care providers, and an appropriate array of scientists and engineers should be strengthened, with assistance from developed nations and regional and multilateral organizations.

Inclusion of cultural values and health education in health care delivery is essential to the treatment, control and prevention of disease and the overall promotion of health, but this important integration is often neglected or overlooked. Comparatively heavy reliance is placed on the conventional classroom model of education, and development of health care and health education is left to the health professionals.

Two fundamental philosophies should prevail in health education: 'start where the people are' and 'involve the people' (Green, 1983). Experience has shown that paternalistic approaches and the imposition of decisions upon others are seldom effective. The 1978 Alma-Ata Declaration and the subsequent WHO Global Strategy of Health for All by the Year 2000 gave new life to the concept of health education by designating 'education concerning prevailing health problems and the methods of preventing and controlling them' as the first essential activities in primary health care (WHO, 1978). A WHO Expert Committee report follows on from this (WHO, 1983b), concluding that 'health science and technology can make a real impact *only* if the people themselves become full partners in health protection and promotion' and that health education goals should be integrated into the planning and management of health programmes at all stages. Health for all by the year 2000 does not mean a utopian world in which there would be no sick or disabled. It means a real world in which individuals, families and communities would not only have access to essential health care but would also be better equipped to prevent illness at home, at school, at the place of work, and would be aware of their power to shape a world free from avoidable or man-made disease' (Mahler, 1984).

Priorities for research within developing countries must take into account the most important health problems and the special conditions that may alter the effectiveness or acceptability of health programmes, from oral rehydration to contraceptives to stability of vaccines. More researchers in universities, government agencies and corporations in the developed countries need to be recruited to work on the endemic diseases of the developing world.

Fortunately, breakthroughs in culturing parasites and evidence of scientific phenomena of general interest in these organisms are generating more excitement in the research community.

Both for preventive services and for care of patients, countries must rely upon the primary health care model advocated by WHO. The concept of essential drugs, the use of oral rehydration therapy and of contraceptives, and the devolution of responsibility and authority within a well-ordered referral system are key elements. An adequate infrastructure for supplies, equipment, training, data collection and inter-ministry planning is essential. An international mechanism for sharing information about the efficacy, safety, management requirements and cost effectiveness of existing and emerging technologies, together with technical assistance for country-specific technology assessment, would be desirable. WHO, PAHO and other international organizations could develop or expand such efforts in health technology assessment.

References

BANTA, D. 1984. The Use of Modern Technologies: Problems and Perspectives for Industrialized and Developing Countries. *Bulletin of the Pan American Health Organization*, Vol. 18, pp. 139-50.

BARRETT, D.E.; RADKE-YARROW, M.; KLEIN, R.E. 1982. Chronic Malnutrition and Child Behavior: Effects of Early Caloric Supplementation on Social-emotional Levels at School Age. *Developmental Psychology*, Vol. 18, pp. 541-56.

BEASLEY, R.P.; HWANG, L.U.; LIN, C.C.; CHIEN, C.S. 1981. Hepatocellular Carcinoma and Hepatitis B Virus. A Prospective Study of 22,707 Men in Taiwan. *Lancet*, Vol. 2, pp. 1129-33.

BELL, D.E. 1980. Introduction: Health and Population in Developing Countries. *Social Science and Medicine*, Vol. 14C, pp. 63-5.

BELL, J.N. 1984. Collaborative Research Systems: The BOSTID Experience in Health to Date. Paper prepared for the *NCIH Conference on Universities and International Health*, University of North Carolina, 23-6 September.

BOURNE, P.G. 1984. *Global Water for Life*. Washington, D.C., Global Water Inc.

BRENNER, A. 1982. Recent Developments in the Field of Chagas' Disease. *Bulletin of the World Health Organization*, Vol. 60, pp. 463-73.

BRINK, R.A.; DENSMORE, J.W.; HILL, G.A. 1977. Soil Deterioration and the Demand for Food. *Science*, Vol. 197, pp. 625-30.

BURNS PARLATO, M. 1982. *Primary Health Care: An Analysis of 52 AID-assisted Projects*. Washington, D.C., American Public Health Association.

CALDWELL, J.C. 1979. Education as a Factor in Mortality Decline: An Examination of Nigerian Data. *Proceedings of the Meeting on Socioeconomic Determinants and the Consequences of Mortality*, pp. 172-92. Geneva, World Health Organization.

CARLOS, J.P. (ed.). 1973. *Prevention and Oral Health*. Conference sponsored by Fogarty International Center for Advances Studies in the Health Sciences and the National Institute of Dental Research, DHEW-NIH 74-707, pp. 9-10.

CHANDLER, W.U. 1984. Improving World Health: A Least Cost Strategy. *Worldwatch Paper* 59. Washington, D.C., Worldwatch Institute.

CHAPARAS, S.D. 1982. Immunity in Tuberculosis. *Bulletin of the World Health Organization*, Vol. 60, pp. 447-62.

CHAPIN, G.; WASSERMAN, R. 1981. Agricultural Production and Malaria Resurgence in Central America and India. *Nature*, Vol. 293, pp. 181-5.

CROMPTON, D.W.T.; NESHEIM, M.C. 1984. Malnutrition's Insidious Partner. *World Health* (March), pp. 18-21.
DAME, D.; WILLIAMS, J.L.; MCCUTHAN, T.F. et al. 1984. Structure of the Gene Encoding the Immunodominant Surface Antigen on the Sporozoite of the Human Malaria Parasite *Plasmodium falciparum*. *Science*, Vol. 225, pp. 593-9.
DEINHARDT, F.; GUST, I.D. 1982. Viral Hepatitis. *Bulletin of the World Health Organization*, Vol. 60, pp. 661-91.
ENVIRONMENTAL SANITATION INFORMATION CENTER. 1983. *Human Waste Management for Low-income Settlements*. Bangkok.
FLEURET, P.; FLEURET, A. 1983. Socio-economic Determinants of Child Nutrition in Taita, Kenya: A Call for Discussion. *Culture and Agriculture*, Vol. 19, pp. 16-29.
GAISE, S.K. 1979. Some Aspects of Socioeconomic Determinants of Mortality in Tropical Africa. *Proceedings of the Meeting on Socioeconomic Determinants and Consequences of Mortality*, pp. 193-207. New York/Geneva, United Nations and World Health Organization.
GALAZKA, A.M.; LAUER, B.A.; HENDERSON, R.H.; KEJA, J. 1984. Indications and Contraindications for Vaccines Used in the Expanded Programme on Immunization. *Bulletin of the World Health Organization*, Vol. 62, pp. 357-66.
GHULAAM MOSTAFA, A.B.M. 1984. Bangladesh: The Nettle Grasped. *World Health* (July), pp. 6-9.
GREEN, L.W. 1983. New Policies in Education for Health. *World Health* (April-May), pp. 12-18.
GREEP, J.M.; SCHMIDT, H.G. 1984. The Network. *World Health* (April), pp. 918-21.
GRIFFITH, D.H.; RAMAR, D.V.; MASHAAL. 1971. Contributions of Health to Development. *International Journal of Health Services*, Vol. 1, pp. 253-70.
HACKMAN, E.; EMANUEL, I. et al. 1983. Maternal Birth Weight and Subsequent Pregnancy Outcome. *Journal of the American Medical Association*, Vol. 250, pp. 2016-19.
HUNTER, J.; REY, L.; SCOTT, D. 1983. Man-made Lakes — Man-made Diseases. *World Health Forum* 3(2).
INSTITUTE OF MEDICINE. 1978. *Strengthening U.S. Programs in Developing Countries*. Washington, D.C., National Academy of Sciences.
–. 1980. *U.S. Participation in Clinical Research in Developing Countries*. Washington, D.C., National Academy of Sciences.
KALBERMATTEN, J.M. et al. 1982. *Appropriate Sanitation Alternatives*. Baltimore, The Johns Hopkins University Press.
KLEIN, R.E. et al. 1976. Effects of Maternal Nutrition on Fetal Growth and Infant Development. *PAHO Bulletin*, Vol. 10, pp. 301-16.
KOLATA, G. 1984. Scrutinizing Sleeping Sickness. *Science*, Vol. 226, pp. 956-9.
Lancet. 1978. Water with Sugar and Salt. *Lancet*, No. 2, pp. 300-1.
MAHLER, H. 1981. The Meaning of "Health for All by the Year 2000". *World Health Forum*, Vol. 2, pp. 5-22.
–. 1984. Population and Health. *World Health* (June), p. 3.
MATA, L.J. 1975. Malnutrition-Infection Interactions in the Tropics. *American Journal of Tropical Medicine*, Vol. 24, pp. 564-74.
MITSUYA, H.; POPOVIC, M.; YARCHOAN, R.; MATSUSHITA, S.; GALLO, R.; BRODER, S. 1984. Suramin Protection of T Cells in vitro against Infectivity and Cytopathic Effect of HTLV-III. *Science*, Vol. 226, pp. 172-4.
MOSLEY, H.W. 1983. *Will Primary Health Care Reduce Infant and Child Mortality? A Critique of Some Current Strategies with Special Reference to Africa and Asia*. Paris, International Union for the Scientific Study of Population Seminar on Social Policy, Health Policy and Mortality Prospects (February).
MUKERJEE, D. 1984. Blighted Cities. *World Health* (June), p. 10.
NATIONAL ACADEMY OF SCIENCES. 1977. *World Food and Nutrition Study*. Washington, D.C.
NATIONAL CENTER FOR HEALTH STATISTICS. 1982. *Health: United States, 1982*. Washington, D.C., United States Government Printing Office.
NATIONAL RESEARCH COUNCIL. 1979. *U.S. Science and Technology for Development: A Contribution to the 1979 U.N. Conference*. Washington, D.C., Department of State.

–. 1982. *Priorities in Biotechnology Research for International Development*. Washington, D.C., National Academy Press.
NAVARRO, V. 1984. Policies on Exportation of Hazardous Substances in Western Developed Countries. *New England Journal of Medicine*, Vol. 311, pp. 546-8.
NIGHTINGALE, S.L. 1984. Essential Drugs, A Guide Through the Maze. *World Health* (July), pp. 26-7.
NORDEEN, S.K.; SANSARRICO, N. 1984. Immunization Against Leprosy: Progress and Prospects. *Bulletin of the World Health Organization*, Vol. 62, pp. 1-6.
OFOSU-AMAAH, V. 1983. National Experience in the Use of Community Health Workers. *WHO Offset Publication* No. 74. Geneva.
OMENN, G.S.; MOTULSKY, A.G. 1978. Eco-genetics: Genetic Variation in Susceptibility to Environmental Agents. In: B.H. Cohen, A.M. Lilienfield, P.C. Huang (eds.), *Genetic Issues in Public Health and Medicine*, pp. 83-111. Springfield, Illinois, C.C. Thomas.
OMENN, G.S.; HOLLAENDER, A. (eds.). 1984. *Genetic Control of Environmental Pollutants*. New York, Plenum Press.
PARKIN, D.M.; STJERNSWARD, J.; MUIR, C.S. 1984. Estimates of the Worldwide Frequency of Twelve Major Cancers. *Bulletin of the World Health Organization*, Vol. 62, pp. 163-82.
PATH (Program for Appropriate Technology in Health). 1983. *Health Technology Directions*, Vol. 3, p. 3.
PATHAK, B. 1982. *Sulabh Shauchalaya: Hand Flush Water Seal Latrine, A Single Idea that Worked*. Calcutta, Amola Prakashan, Patna.
PETROS-BARVAZIAN, A. 1984. Family Planning: A Preventive Health Measure. *World Health* (June), pp. 4-7.
PRENTICE, R.L.; OMENN, G.S. et al. 1984. Rationale and Design of Cancer Chemoprevention Studies in Seattle. Fourth Symposium on Epidemiology and Cancer Registries in the Pacific Basin, Kona, Hawaii, January.
ROCKEFELLER FOUNDATION. 1984. *Protecting the World's Children: Vaccines and Immunization*. New York.
SCRIMSHAW, N.S.; TAYLOR, C.E.; GORDON, J.E. 1968. Interaction of Nutrition and Infection. *World Health Organization Monograph* No. 57, Geneva.
SCUDDER, T. 1972. Resettlement. In: *Man-made Lakes as Modified Ecosystems* (SCOPE Report 2), pp. 451-71. Paris, International Council of Scientific Unions.
STANDLEY, T.; KESSLER, A. 1983. Trail Blazing. *World Health* (December), pp. 8-10.
STEENSTRUP, J.E. 1984. Rationing in Rural Kenya. *World Health* (July), pp. 18-19.
STEWART, G. 1984. Education Behavioral Sciences Research Needs and Approaches. Discussion Outline for the *NICH Conference on Universities and International Health*, University of North Carolina, 23-6 September.
THOMAS, L. 1983. *The Youngest Science*. New York, Viking Press.
–. 1984. Scientific Frontiers and National Frontiers. *Foreign Affairs* (Spring).
TULI, J. 1984. Drink of Life. *World Health* (January-February), pp. 914-15.
UNDERWOOD, B.A.; TUPULE, P.G. 1984. *NEI-India Collaborative Research on Prevention of Blindness*. Scientific seminar presented to the Nutrition Coordinating Committee, Bethesda, Maryland, 10 May.
US-AID (United States, Agency for International Development). 1982. *Safe Water and Waste Disposal for Human Health: A Program Guide*. Washington, D.C.
VESIKARI, T.; ISOKURI, E.; DELEM.; D'HONDT, E.; ANDRE, F.E.; ZIZISS, G. 1983. Immunogenicity and Safety of Live Oral Attenuated Bovine Rotavirus Vaccine Strain RIT 4237 in Adults and Young Children. *Lancet*, No. 2, pp. 807-11.
WALDEMAR, F.A. 1984. The Dangers and the Precautions. *World Health* (August-September), pp. 10-12.
WALSH, J.A.; WARREN, K.S. 1979. Selective Primary Health Care. An Interim Strategy for Disease Control in Developing Countries. *New England Journal of Medicine*, Vol. 301, pp. 967-74.
WARREN, K.S. 1980. The Relevance of Schistosomiasis. *New England Journal of Medicine*, Vol. 303, pp. 203-6.

WARREN, K.S.; BOWERS, J.Z. (eds.). 1983. Parasitology, A Global Perspective. New York, Springer-Verlag. (Note esp. Davis, pp. 62-74; Lucas, pp. 256-69; Harinasuta, pp. 19-44).
WINICK, M. 1976. *Malnutrition and Brain Development*. New York, Wiley.
WORLD BANK. 1975a. *Health Sector Policy Paper*. Washington, D.C.
–. 1975b. *Internal Migration in Less Developed Countries: A Survey of the Literature*. Washington, D.C.
World Health. 1984a. Injuries at work. *World Health* (May), p.31.
–. 1984b. Special Report: Cancer is a Third World Problem, Too. *World Health* (June) p. 30.
WORLD HEALTH ORGANIZATION (WHO). 1975. *Report of the Health Situation, 1969-1972*. Geneva.
–. 1977. The Selection of Essential Drugs. WHO Technical Report Series, No. 615. Geneva
–. 1978. *Primary Health Care: Report of the International Conference on Primary Health Care, Alma-Ata*. Geneva.
–. 1979a. Cancer Statistics. WHO Technical Report Series, No. 632. Geneva.
–. 1979b. Controlling the Smoking Epidemic. WHO Technical Report Series, No. 636. Geneva.
–. 1979c. The Selection of Essential Drugs. WHO Technical Report Series, No. 641. Geneva.
–. 1980a. Epidemiology and Control of Schistosomiasis. WHO Technical Report Series, No. 643. Geneva.
–. 1980b. Resistance of Vectors of Diseases to Pesticides. WHO Technical Report Series, No. 655. Geneva.
–. 1980c. *Sixth Report on the World Health Situation 1973-1977. Part I: Global Analysis. Part II: Review by Country and Area*. Geneva.
–. 1981. *Rapid Laboratory Techniques for the Diagnosis of Viral Infections*. WHO Technical Report Series, No. 661. Geneva.
–. 1982a. *Tuberculosis Control*. WHO Technical Report Series, No. 671. Geneva.
–. 1982b. *Prevention of Coronary Heart Disease*. WHO Technical Report Series, No. 678. Geneva.
–. 1983a. *The Use of Essential Drugs*. WHO Technical Report Series, No. 685. Geneva.
–. 1983b. *New Approaches to Health Education in Primary Care*. WHO Technical Report Series, No. 690. Geneva.
–. 1983c. *Viral Vaccines and Antiviral Drugs*. WHO Technical Report Series, No. 693. Geneva.
–. 1983d. *Smoking Control Strategies in the Developing Countries*. WHO Technical Report Series No. 695. Geneva.
–. 1984a. *Mental Health Care in Developing Countries: A Critical Appraisal of Research Findings*. WHO Technical Report Series, No. 698. Geneva.
–. 1984b. *Leishmaniases*. WHO Technical Report Series, No. 701. Geneva.
YOSHIDA, N.; NUSSENZWEIG, R.S et al. 1980. *Protective Antibodies Directed Against the Sporozoite Stages of Malaria Parasite*. *Science*, Vol. 207, pp. 71-3.
YOUNG, D.B.; BUCHANAN, T.M. 1983. A Serological Test for Leprosy with a Glycolipid Specific for *Mycobacterium leprae*. *Science*, Vol. 221, pp. 1057-9.

5

New energy technologies and systems

Energy resources and problems

The production and use of energy have always been closely linked with the evolution of human society: all major changes in the organization of society, from the Bronze Age onwards, have been accompanied by the exploitation of new energy sources and technologies. In the twentieth century, oil and natural gas as energy sources, electricity as a method of distributing energy easily, and the internal combustion engine (in road vehicles and aeroplanes) have made possible new lifestyles characterized by high mobility of both persons and goods, and by the widespread diffusion of information. However, relatively few human societies have fully benefited from these changes, with the result that there are extraordinary differences in lifestyles today. These differences can be vividly illustrated by the variations in energy consumption per capita: citizens of the most developed countries of North America consume over twenty times as much energy on average as those of the least developed countries of the world.

In recent years, the basis of much of the world's energy system has been buffeted by shocks. For the more developed countries, special importance has been ascribed to the two sharp increases in the price of oil which took place in the 1970s. But in both the developed and developing world alike, a more insidious threat comes from the possibility that traditional energy sources are fast becoming depleted. This applies to the hydrocarbons, but equally and more dramatically to firewood and kindling and even to dung in the non-commercial energy sectors of the poorest subsistence economies.

Energy systems in developed countries have recently undergone a complex evolution in response to the crisis 1970s, and many new energy technologies have been proposed, developed and criticized. An economic crisis, whose causes and issues are not yet completely clear, affected all countries after the close of the 1970s. Developing countries have seen growth rates slow down considerably, leading to rethinking of a model of development hitherto held to be without alternative: it had been taken for granted that development policies inevitably entailed large increases in energy consumption. Furthermore, there was a belief that energy technologies would be able to respond to increases in demand without producing unacceptable strains or side-effects. The corollary of this was the notion that developing countries would have an energy evolution parallel to that of the developed countries. The whole question has now been reopened. Less uniform views are developing, based on analytical examination of energy problems, the various technologies available to resolve them, and their interactions with the social context and with the environment.

EXPERIENCES SINCE 1973

The world's current energy problems derive from the fact that in the 1970s an international economy based on low-cost energy experienced sudden increases in the price of its main energy source, oil. Between 1973 and today, there was approximately a fourfold price increase in real terms, with consequences of the greatest importance since previously oil had followed in seemingly limitless quantities and at a price which undermined the use of all other potential alternative sources. To compound these problems, oil imports came increasingly from a historically unstable geopolitical area; interruptions of supplies were therefore a real possibility, and the mere rumour of impending conflict could disturb the oil market severely.

Industrialized countries

The effects of the oil price increases have been different in the developed and developing countries. In the former, new energy technologies to tackle the changed situation were already available or could be developed to diminish reliance on oil: advances in mining and combustion techniques contributed to a revival of the use of steam coal as a major energy source, while thermal nuclear reactors were able to assume an essential role in the provision of electricity, especially for base load. Progress in oil exploration and production technologies has made it possible to exploit new deposits even in environments previously regarded as too hostile, such as sub-arctic areas and the open sea. Technologies relating to the transport of energy over long distances enabled new forms of connection between primary energy sources and areas where demand for energy is greatest.

In many human activities, including essential industrial processes, the use of known but hitherto neglected technologies permitted a significant reduction in energy consumption; the availability of complex supervision and control systems at low cost owing to advances in the various branches of microelectronics (computers, information technologies, automation and robotization) has made such savings even greater. Introduction and application of these technologies, however, took longer than had originally been thought, because of the inherent complexity of the energy system in a developed country, the time required to construct large plants, the large capital requirements, and the need to educate public opinion in the new energy priorities.

At the same time, industry in the developed countries was compelled to reorganize — a process induced or accelerated by high energy prices. The current trend is away from standardized mass-produced goods, toward high-quality, customized goods and services in which the decisive input is technological, while the energy content per unit of value added is relatively low. A similar quantity of goods and services, of higher quality, can therefore be obtained using less raw materials and less energy.

The exploitation of such technological opportunities allowed advanced countries to contain their overall energy consumption (in 1983 actually a few percent lower than in 1973), to reduce their dependence on oil from 53.3% to 44.6% — and their dependence on Middle Eastern oil even more (Table 1). Their GNPs nevertheless rose by about 20% in real terms. These figures mask differences — a greater-than-average reduction in energy consumption was achieved in the industrial sector (no less than 49% in Japan, 30% in the leading countries of Western Europe and 20% in the United States of America) with a lower reduction in energy consumption for domestic purposes (about 18% for OECD countries as a whole), and still less for transport (about 11.5%). As oil diminished in importance, electricity as an energy vector continued to grow in Western industrialized countries during the decade after the first energy crisis. Electricity's share, calculated as a primary source itself rather than as a percentage on end uses, increased from 25% to about 32%, thus confirming that its flexibility and ease of use make it particularly suited to the demands of modern life (Colombo, 1982).

Table 1. Energy consumption in Western industrialized countries. (million tonnes oil equivalent)

	Oil	Coal	Natural gas	Hydro-geo	Nuclear	Total
1960	764 (40.1%)	663 (34.8%)	318 (16.7%)	158 (8.3%)	1 (0.05%)	1904
1973	1955 (53.3%)	681 (18.55%)	753 (20.5%)	236 (6.45%)	44 (1.2%)	3669
1979	1985 (50.75%)	739 (18.9%)	784 (20.05%)	271 (6.95%)	133 (3.4%)	3912
1982	1625 (45.2%)	783 (21.75%)	722 (20.05%)	282 (7.85%)	184 (5.1%)	3596
1983	1592 (44.6%)	786 (22.0%)	701 (19.65%)	291 (8.15%)	200 (5.6%)	3570

Source: ENEA (Rome) based on *British Petroleum Statistical Review of World Energy*.

The most significant changes in energy indices of the advanced industrial countries took place in the five years from 1979 to 1983. These economies suffered wide fluctuations during the same period, so that it is not easy to measure accurately the global effects of the energy crisis and the adoption of new technologies on the economy and society. It is, however, clear that an irreversible change is under way as regards the model of development which was at the base of the protracted economic boom of 1950 to 1970. This model appears, in fact, to have lost relevance in developed countries while showing limited potential for implementation in developing countries.

The new trends seem to have broken the close link hitherto existing between development and growth in energy consumption, and to have put an end to the predominance of one single energy source over all others, in favour of a more composite and flexible energy system. The consequences of this change for the economy, employment, lifestyle, social and political structure, and indeed value systems, may be clearer in the future.

Centrally planned economies

In the USSR and other Eastern European countries, the energy situation in 1973 was markedly different. The share of oil in the energy supply system was only 32.5%, the predominant energy source being coal (Table 2). The effects of the energy crises permeated more slowly, given that the centrally planned economies have an internally self-sufficient energy balance. It was only later that demand and supply patterns within this group began to reflect changes in the world scenario so that, as a consequence of the altered international terms of trade, these countries too experienced energy supply difficulties. Though therefore delayed, reverberations proved not dissimilar in their general characteristics to those experienced by developed Western nations.

Table 2. Energy consumption in countries with centrally-planned economies. (million tonnes oil equivalent)

	Oil	Coal	Natural gas	Hydro-geo	Nuclear	Total
1960	150 (15.5%)	737 (76.2%)	46 (4.75%)	34 (3.5%)	—	967
1973	468 (28.6%)	870 (53.15%)	247 (15.1%)	48 (2.9%)	4 (0.25%)	1637
1979	636 (29.35%)	1064 (49.1%)	381 (17.6%)	70 (3.2%)	17 (0.8%)	2168
1982	651 (28.5%)	1064 (46.55%)	460 (20.1%)	83 (3.65%)	28 (1.2%)	2286
1983	653 (27.7%)	1098 (46.6%)	490 (20.8%)	87 (3.7%)	29 (1.25%)	2357

Source: ENEA (Rome) based on *British Petroleum Statistical Review of World Energy*.

Other difficulties have recently been encountered in reaching all the objectives set in national and Comecon plans for the energy sector. These objectives, in their general lines, parallel those pursued in the West — though greater emphasis is placed on programmed energy exchanges and on the joint exploitation of the massive primary energy resources of the various countries, most especially the USSR. The trend for electricity to increase its share in energy distribution follows the world pattern: its share already rose from 20% to about 25% between 1973 and 1983.

Developing world

In developing countries, energy consumption has continued to grow even after 1973 (Table 3), to meet the demands imposed by high rates of population growth and efforts to raise low standards of living. The increased cost of imported energy supplies has thus led to a reduction in the growth rate of real gross domestic product — in the four years 1979-83, 40% lower than in the period before 1973. The actual rate of increase in energy consumption fell rather more during the same period, by about 50%. Even to achieve this minimum performance, developing countries as a whole have had to borrow massively. Their debts have become a source of instability for the international economy and not merely an obstacle to development for the countries themselves.

Table 3. Energy consumption in developing countries.
(million tonnes oil equivalent)

	Oil	Coal	Natural gas	Hydro-geo	Nuclear	Total
1960	130 (53.05%)	80 (32.65%)	15 (6.1%)	20 (8.15%)	—	245
1973	375 (61.8%)	117 (19.3%)	66 (10.85%)	48 (7.9%)	1 (0,15%)	607
1979	503 (58.5%)	173 (20.1%)	109 (12.65%)	72 (8.35%)	3 (0.35%)	860
1982	549 (56.25%)	200 (20.5%)	135 (13.85%)	86 (8.8%)	6 (0.6%)	976
1983	549 (55.0%)	213 (21.35%)	138 (13.85%)	91 (9.1%)	7 (0.7%)	998

Source: ENEA (Rome) based on *British Petroleum Statistical Review of World Energy*.

The developing countries, too, have switched their sources of supply of energy between 1973 and 1983, reducing dependence on oil from about 62% to 55%. The role of electricity has continued to increase here as well, rising from about 15% of commercial energy consumption in 1973 to around 20% in 1983.

Energy conservation technologies, the diversification of sources, and structural changes in industry and the service sector in developed countries have contributed to stabilizing the price of oil and of energy generally, thus restoring one of the preconditions for a resumption of economic growth. A more optimistic economic climate could eventually improve the situation of the non-oil producing countries of the developing world, expanding markets for their raw materials and manufactures; the debt burden may then be lightened. Oil producers, many of whom are also developing countries, benefit from more stable prices for their chief export, from more orderly markets and stable currencies.

ENERGY CONSUMPTION FORECASTING

The energy crisis called the attention of governments and public opinion to the serious energy problems that humanity is likely to face eventually. Energy systems based on an assumption of universally available, reasonably priced energy resources are no longer going to be the norm. Furthermore, and in the longer term, ecological and ethical considerations regarding our right to continue the depletion of natural resources to the detriment of generations to come are going to be equally impelling. The problem that energy technologies must therefore solve is to provide over time an ever greater number of human beings (perhaps twice as many as now populate the Earth, according to recent estimates) with the energy required to provide a satisfactory standard of living (Freeman and Jahoda, 1978; Colombo and Bernardini, 1979).

We do not yet possess an energy technology tested on a sufficiently large scale that is capable of guaranteeing such a level of energy consumption, if the fossil fuels are excluded. Continued exploitation of these sources implies the assumption of a great responsibility toward future generations, and the parallel commitment to develop alternative technologies independent of them. A very special kind of investment evaluation is required to assess correctly the benefits of such long-term research and development programmes. Ordinary cost-benefit analysis may underestimate the importance of factors such as our paramount need to safeguard energy sources for future generations, our need to be absolutely sure — in so far as this is possible — that only environmentally benign energy options are taken up (Kemp and Long, 1984).

The progress achieved in the technologies of producing fuels and hydrocarbons from coal and now also from natural gas allows us to forecast that future generations will be able to face the gradual exhaustion of oil deposits without dramatic effects on their standard of living. Progress in the technologies of fast breeder reactors, nuclear fusion, photovoltaic solar energy and the production and utilization of hydrogen give us some reassurance regarding the possibility of a more general depletion of non-renewable sources — including the other fossil fuels and even uranium ores. These developments give energy economists reasonable grounds for their forecasts of energy scenarios for the future.

Some models

The application of new technologies and the development of new energy systems are strongly influenced by both quantitative and qualitative trends in energy consumption. To evaluate the prospects for the various technologies, quantitative forecasts are needed of trends in world energy consumption and of consumption at the level of groupings of nations and of individual countries. Considerable effort has recently been dedicated to the construction of energy consumption forecasting models, a problem which was never tackled systematically until the energy crisis revealed the limitations of extrapolation from past trends. Energy — as a conditioning factor of economic growth and development — had previously not been given due weight. After 1973, this situation was rectified to some degree.

Present models can be divided first into partial and total models. Partial models do not examine all the relations between energy and the economy, but concentrate on those assumed to be most important. An example of a partial model is that developed at the International Institute for Applied Systems Analysis in Laxenburg (Marchetti and Nakicenovic, 1979). This applies to the energy sector a technology-substitution model based on the observation of certain irregularities in the use of various energy sources over the last 200 years, and it generates from them the usual logistic curves. According of this model, the use of coal as an energy source is now in an irreversible stage of decline. The energy crisis of 1973-80 is held to coincide with the maximum penetration of oil. Among sources destined to replace oil, the model identifies first natural gas and later nuclear power. Empirical observation of periods of transition from one source to another indicates that they are characterized by wide fluctuations in energy prices, whereas periods in which a new predominant source has established itself are characterized by a return to low-cost energy.

The model has been further elaborated, likening periods of transition from one energy source to another to periods of rapid penetration of key technological innovations in advanced economies with expansion stemming from exploitation of the new source and its related technologies. In a certain sense, therefore, this links through to the long-cycle (Kondratiev) theory of growth in the capitalist world.

The total models can, in turn, be divided into aggregate (macroeconomic) models and disaggregated models. Many aggregate models aim to establish relationships between the price of energy and economic development at the macroeconomic level. Some of them also take into account differences between the energy sources in relation to end uses, and therefore permit forecasting not only of the overall trend in the economy and in total energy consumption but also of trends in end use of the various sources.

Disaggregated models divide the economy into a fairly large number of productive sectors, establishing correlations between them and between each sector and the energy sector. They are models of great complexity, requiring vast input of data and expertise (Gass, 1984).

All models are limited by the fact that they can take only economic and technological factors into consideration. There are certain consequences of economic and technological transformation which go beyond the fields of economics and energy, creating social and political strains which can give rise to uncertainties and unpredictable events. This complex of interactions may feed back into the economic and energy systems, causing discontinuities that are difficult to foresee. To overcome these problems, hypotheses or variables generated outside the model are introduced, but inevitably these may reflect the expectations of the person selecting them. Many models, particularly the more complex, may be especially sensitive to these hypotheses and variables; small differences in them may lead to very large differences in conclusions over the long run (Keepin, 1984).

Estimates of energy consumption trends in the medium term tend nowadays to be made by building a range of scenarios, generally differing in the assumed rate of growth of the economy (high, medium and low) and in the predicted share taken by oil (and therefore in the forecast trend for the relative price of oil). The models are then applied to determine total energy consumption in each case, with the shares of other energy sources and all the other variables assumed dependent (Chateau, 1985). In this way, intervals and most-likely values are determined, and the possibilities of application of new energy technologies are evaluated on this probabilistic basis.

Models can also have a certain utility in estimating the short-term economic consequences of various energy policies or, *vice versa*, the energy consequences of economic policies. They have become an indispensable analytical tool for those involved in energy decision-making.

New energy technologies

New energy technologies are emerging which may soon offer wider horizons leading to the satisfaction of mankind's energy needs.

OIL

Oil is still the world's major energy source. It is predominant in international energy commerce, having always been traded widely: 54% of oil produced in 1974 was exported, falling to 43% in 1983; corresponding values are 12% for natural gas and 5% for coal. Oil as an energy source is for many uses indispensable to our way of life.

Oil technologies are now well proven, though the energy crisis stimulated considerable development even in these (Hobson, 1983). The need to explore for oil at ever greater depths and in increasingly hostile environments has greatly increased the cost of exploratory drilling. It is thus important to exploit all geophysical techniques available to decrease the incidence of dry wells. Advances in instrumentation, data collection and mathematical methods of data reduction and computer processing have expanded enormously the extent of knowledge of what is under the surface before drilling starts. Cost savings have been remarkable, especially considering the amount of new exploration called for after the oil crisis.

Table 4. Oil production.
(million tonnes)

	Western industrialized countries	Centrally planned economies	Oil exporting developing countries	Other developing countries	World total
1973	644.5 (22.45%)	503.1 (17.5%)	1580.6 (55.05%)	143.4 (5.0%)	2871.6 (100%)
1976	585.0 (19.8%)	623.4 (21.1%)	1595.1 (54.0%)	150.8 (5.1%)	2954.3 (100%)
1979	681.7 (21.15%)	712.2 (22.05%)	1639.3 (50.8%)	193.4 (6.0%)	3226.6 (100%)
1982	705.6 (25.35%)	734.2 (26.35%)	1132.6 (40.65%)	212.3 (7.6%)	2784.7 (100%)
1983	729.2 (26.45%)	741.9 (26.9%)	1052.0 (38.2%)	232.4 (8.45%)	2755.5 (100%)

Source: ENEA (Rome) based on *British Petroleum Statistical Review of World Energy*.

In exploration techniques, important progress has been made in deep drilling and in drilling at sea. Deep drilling, with bit temperatures which may reach 250-300°C, has required the development of new drilling fluids and cementing materials able to resist such temperatures. The technology of drilling at sea is of recent origin, beginning in the 1940s when operations started in just a few metres of water. Depths has reached 350 m in 1972, and now go down to almost 2,000 m (Table 5). In deep-water drilling, use is made of floating platforms or ships held in position by computer-controlled thruster motors. Off-shore production platforms can now be steel structures as high as skyscrapers, with weights as great as 600,000 tons. To resist strains and stresses imposed by movements of the seabed and buffeting by wind and wave, these structures have to be designed using computer-aided design (CAD) techniques, with their behaviour under static and dynamic loads simulated on a computer. To ensure maximum resistance to fatigue, quality control techniques previously developed for the nuclear and aerospace industries are adopted.

Table 5. Maximum water depth in off-shore oil exploratory drilling and production wells.
(Water depth in metres)

	Exploratory drilling	Production well
1947	7.5	6.0
1965	192.0	87.0
1969	319.0	104.0
1976	1055.0	260.5
1978	1324.5	312.5
1983	1965.0	312.5

Source: Petroski, 1984.

Off-shore oil production technology also received strong impetus from the energy crisis. In 1972, the maximum workable depth of water was a little over 100 m; over 300 m has now been exceeded and even greater depths will be achieved shortly (Table 5). The number of off-shore production wells has risen from a few hundreds to many thousands, and oil produced as a proportion of total output from one sixth to a quarter. To reach still greater depths, fully submerged (sub-sea) production systems are being developed, along with lighter towers (e.g. guyed towers and tension-leg platforms). Wells have recently also been drilled under the pack-ice in the Barents and Beaufort Seas, relying on progress in cryoresistant materials and techniques. Off-shore fields are likely to increase their contribution to total oil output in the future.

Problems of international politics have already arisen in many areas – between the Socialist People's Libyan Arab Jamahiriya and Malta in the Mediterranean, the Islamic Republic of Iran and Sharjah in the Persian Gulf, to name but two examples — as international law on the exploration and exploitation of the continental shelf and seabed is unclear. The UNEP Regional Seas Programme is already attempting to confront the problem as regards two of its eight areas of concern — the Mediterranean and the Caribbean. It is then intended to use experience gained in these regional fora for the other designated regional seas: the Red Sea and Gulf of Aden, the Kuwait Action Plan — Gulf Region; the South-East Pacific, the West African, the East Asian Seas and the South West Pacific. But the whole wider question of the international Law of the Sea is a vexed one, and the December 1982 signing ceremony of the United Nations Law of the Sea Convention in Montego Bay, Jamaica, has not resolved matters as regards under-sea prospecting for and exploitation of energy sources. Until the advanced industrial economies are convinced of the rightness of solutions proposed there, more especially of the establishment of the supranational body Enterprise as effective world holder of under-ocean rights, then there will continue to be uncertainties limiting investment decisions. The regime adopted by majority vote against the wishes of the industrial economies of both East and West at Montego Bay — though reflecting the priorities of the New World Economic Order in its decision to give first claim to protection of the rights of states at present at a stage of development not sufficiently advanced to take immediate advantage of the new under-ocean exploration and exploitation techniques — is virtually unworkable as it now stands. To some extent, the disputes which have arisen within the United States of America between federal and state governments constitute a significant precedent, hampering full exploitation of the productive potential of the North American outer continental shelf (Lawrence, 1984).

The increase in the price of crude oil has also encouraged progress in technologies to increase the percentage extraction of oil from wells. The proportion of oil in a field which flows out spontaneously varies, but before 1973 the only economically viable techniques for increasing this fraction were the injection of water or the re-injection of part of the natural gas generally associated with the oil. Such techniques allow the recovery rate to be raised somewhat, but it is now economic to use more costly methods to increase the recovery rate still further. Methods currently under trial include injection of hot liquids, injection of gas (generally carbon dioxide or nitrogen), and injection of chemicals such as surfactants.

The demand pattern for refinery products has also changed as a result of the increased price of crude, mainly reducing demand for heavy crude. Conservation programmes have affected certain of these products more than others. At present prices, the alternatives (especially coal, but now also natural gas) replace the heavier products of oil distillation. In electricity generation, nuclear energy has taken up a consistent share of demand. Energy conservation — most effective in the industrial sector — has particularly reduced the demand for heavy products. The wider use of the diesel engine has also altered the balance of the demand for oil products. However, the use of refining residues as an energy source for the production of heat and electricity is often hampered by environmental protection legislation, as the concentration of heavy metals and sulphur in these residues — especially in those produced in ever more widespread cracking processes — is often far too high.

Increased oil prices have therefore caused a marked alteration in the break-down of demand between the various fractions produced in the distillation of oil, in favour of the medium and light fractions. Greater use of refinery processes that raise the yield of medium and light products while reducing the impurities in refinery residues has been rendered necessary. Excess refining capacity in developed countries, and the expressed intention of producing countries to increase the value added of their output by exporting refined products instead of crude, are leading to a radical restructuring of the world oil refining industry that also involves the petrochemical feedstocks. There are many processes already available, and others in the development stage, to permit the production, according to market demand, of a variety of refined products from crudes of any composition.

Following on from advances in production technologies and refining, and the more rational use of oil products, recent estimates allow us room for cautious optimism. According to these estimates, in the year 2000 proven oil reserves extractable at a cost below current prices will still stand at a level comparable to the present — i.e. twenty-five to thirty years' consumption. This forecast is, however, conditional upon the present oversupply of oil on the market not leading to a reduction in prospecting, the bringing of new fields into production, and the more rational use of oil derivatives (Desprairies et al., 1984).

These reserve figures themselves do not take into account the possibility that advances in technology could eliminate the commercial penalty at present exacted for the extraction of liquid oil from the vast deposits of tar sands, asphalts and oil shales, extensive reserves of which are to be found in Canada, Venezuela, the United States of America and the USSR. Extractable reserves of fossil hydrocarbons would be multiplied by a factor of at least three should we be able to include the high molecular weight hydrocarbons contained in low concentrations in these rocks and sands. Many different processes for the mining and the production of crude oil from tar sand and oil shale have been proposed or are under investigation, and pilot plants have been in operation for some time. All such processes share similar disadvantages and difficulties — economic, logistical and environmental. The yield in oil per ton of rock or sand is low, less than half a barrel. Mines cover vast areas with very high investment costs. All processes require huge quantities of water and produce enormous amounts of polluting wastes. Likely large-scale production costs of crude oil remain uncertain: cost estimates made some years ago ranged from US $25 to $40 per barrel (Rider, 1981), but they should probably now be revised upwards, further postponing the time when these non-conventional hydrocarbons become fully competitive.

NATURAL GAS

Natural gas possesses the inestimable advantage of being a very clean and flexible source of energy, with uses in industry, the home and in the generation of electricity. Proven reserves are abundant and now equal more than fifty years' supply at the present rate of consumption. They have doubled in the last ten years, and it is forecast that they will continue to increase at the same rate for some decades (Davis, 1984). Some energy analysts therefore see natural gas as the energy source destined to become preponderant in the near

Table 6. Natural gas production. (million tonnes oil equivalent)

	Western industrialized countries	Centrally planned economies	Developing countries	World total
1973	744.3 (67.15%)	260.4 (23.5%)	103.7 (9.35%)	1108.4 (100%)
1976	706.2 (60.05%)	345.9 (29.4%)	123.7 (10.5%)	1175.8 (100%)
1979	741.9 (55.4%)	412.3 (30.8%)	184.8 (13.8%)	1339.0 (100%)
1982	663.9 (48.6%)	507.8 (37.15%)	194.8 (14.25%)	1366.5 (100%)
1983	615.5 (45.65%)	538.4 (39.9%)	194.6 (14.45%)	1348.5 (100%)

Source: ENEA (Rome) based on *British Petroleum Statistical Review of World Energy*.

future (Marchetti and Nakicenovic, 1979). Nevertheless, growth in production of natural gas — extremely rapid between 1960 and 1970 — slowed down in the decade after the oil crisis (Table 6). A principal reason is the fact that the large natural gas fields still to be exploited (whether or not these are associated with oil) are located far from the main centres of demand.

Long-distance transport of natural gas may be either by liquefaction and then low-temperature transport by road, rail or sea, or else by gas pipeline. Liquefaction of natural gas reduces its volume by about 600 times; the process is used for transportation, and also for storage in tank farms or underground caverns. Large-scale commerce in liquid natural gas (LNG) started about 1960, but the development of gas liquefaction technologies and the construction of fleets of LNG carriers proved to be a longer and more difficult process than at first envisaged, bedevilled by design errors, unexpected gas losses during liquefaction, transport and storage, and accidents — some of them fatal. The cost of transporting LNG thus became economic only for very long distances and deep-sea crossings. Liquefaction plants at the gas fields constitute very costly and rigid infrastructures, imposing almost as great a constraint as gas pipelines. Thus, only a fifth of the international trade in natural gas today is in LNG, and three quarters of that is headed for Japan.

Development of gas pipelines has proceeded more regularly, made possible by the perfection of technologies for the manufacture and installation of large-diameter steel pipes resistant to relatively high pressures and low temperatures, by the development of compressors of adequate size, and process computers able to regulate in real time the dynamics of systems now often extending across many thousands of kilometres.

The proportion of natural gas which enters the international market is a mere 12% of the total output, a figure which has grown considerably from the 9% of 1973. Most of the gas trade by sea in liquefied form is between the Gulf States and the South-East Asian producers and Japan, and between North Africa and Western Europe. Algeria is now connected to Italy via an undersea pipeline which has required highly sophisticated engineering. Inside Comecon, a natural gas pipeline network has functioned for some years. New advances in pipeline technology — both in construction and operation — have enabled the USSR to tap the very large reserves of its Siberian fields for export to Central and Western Europe. Such long-distance transport systems require vast investments of capital and labour, and the effort involved is only possible in an international context of long-term guaranteed off-take contracts with a fixed minimum gas price. Financial and political difficulties have delayed exploitation in recent years — these sorts of agreements create very long-standing energy dependences which clearly have a global strategic dimension.

Barriers to natural gas penetration have arisen from differences between producing countries and gas distribution utilities. The producer countries tend to link the price of natural gas to that of oil. The utilities have made large investments in distribution systems, serving users with very varied needs. As long as gas prices were low, distributors were able to a certain extent to play the market, thus maximizing their cost advantage. One of their chief operating concerns is to achieve a load as constant as possible over time, hence they prefer a market with stable, low prices. At times when the oil price rises rapidly, or fluctuates considerably, conflicts of interest arise.

Fuel cells represent a technological development which shows promise for the future in connection with the use of natural gas. From a purely theoretical point of view, fuel cells are very attractive, transforming chemical energy into electricity directly without passing through heat, which allows them to achieve very high efficiencies. The use of natural gas — with its minimal pollution problems — to feed fuel cells would permit the siting of electricity generating plants even within densely populated areas, reducing distribution losses and simplifying co-generation. Fuel cell plants have a modular character, which allows more rapid response to demand and shorter construction times. Utilities could thus reduce their need for risky middle-term forecasting of electricity demand (Langley, 1983; Marshall, 1984).

Prototypes of phosphoric acid fuel cell plants with power outputs of 40-50 kw and 7.5-11 Mw are currently under test. So far, however, the efficiency of these plants has not proved much higher than that of thermal power stations, and capital costs per unit of power are much greater given the limited life of the very expensive electrodes. Hopes of reducing capital cost per unit of power are centred on cells which achieve much higher efficiencies, use less costly electrode materials, have a greater power density per unit of volume and could operate at elevated temperatures (the best examples being molten carbonate cells and solid electrolyte cells). Such cells are as yet at an early stage of development. Given favourable results, the 1990s could see the first introduction of fuel cell plants for power generation (Langley, 1983).

COAL

For a hundred and fifty years — from 1810 to 1960 — coal was the dominant commercial energy source in industrialized countries. At the start of the 1920s, coal provided about 70% of their total energy consumption. Coal deposits are more extensively and evenly distributed than are those of oil or natural gas. This factor, as well as greater practical difficulties involved in transport, has meant that coal has mainly been consumed inside its country of origin. Even in coal's heyday, international trade in the commodity only involved a very modest proportion of output, unlike the situation that arose with oil.

From 1920 onwards, coal's share of world energy consumption declined in favour of oil and natural gas, falling to 40% in 1965 and 28% in 1973. The rise in oil and gas prices then reversed the trend, and coal's share went back to 30.5% in 1983. Given constantly rising global energy consumption, the consumption of coal increased by an average of 4.2% annually between 1860 and 1920, 0.64% between 1920 and 1960 and 2% between 1960 and 1983. Advances in coal technology, substantially reducing the drawbacks, have made a decisive contribution to this turn-round. Other plus factors include security of supply — the wide geographical distribution of coal deposits makes it possible to avoid excessive dependence on a limited number of suppliers — and the size of known reserves, equal to 200 years' demand at current rates of consumption.

Table 7. Coal production.
(million tonnes oil equivalent)

	Western industrialized countries	Centrally planned economies	Developing countries	World total
1973	657.9 (39.45%)	869.2 (52.1%)	141.3 (8.45%)	1668.4 (100%)
1976	665.6 (37.25%)	957.2 (53.55%)	163.9 (9.15%)	1786.7 (100%)
1979	709.1 (35.9%)	1066.7 (54.0%)	200.0 (10.1%)	1975.8 (100%)
1982	749.6 (36.6%)	1064.0 (51.95%)	233.6 (11.4%)	2047.2 (100%)
1983	754.0 (35.95%)	1098.1 (52.35%)	245.0 (11.7%)	2097.1 (100%)

Source: ENEA (Rome) based on *British Petroleum Statistical Review of World Energy*.

The disadvantages derive largely from the fact that coal is a solid (hence more difficult and costly to transport than oil or natural gas) and contains many impurities. It often has to be extracted by deep mining — an unpleasant and dangerous occupation; the difficulties of recruiting and managing a skilled workforce are considerable and tend to increase as the average standard of living in a country rises. This drawback has been reduced by progress in mining technology and automation and in strip-mining techniques, permitting an increase in the percentage of coal obtained from opencast mines, where output per man is greater and the work less disagreeable and dangerous. Output per man in coal mining, opencast and underground, approximately doubled between 1950 and 1970 in the advanced industrialized economies, so that costs have been kept down and coal has remained competitive with oil for some uses — especially for electricity generation (Grainger and Gibson, 1983).

The potential for future major advances in mining technologies seem to lie in two directions — robotization of all face-working (and possibly then all other underground operations) and underground gasification. Experiments in both directions are under way, and advanced automated mines — the first step towards full robotization — are already operating in Great Britain and elsewhere in the industrial West. In the future, telechiric mining techniques — that is, completely automated mining whereby life-support systems are no longer required underground as human beings do not have to work below the surface — could revolutionize coal production in the future. The economics of marginal pits could be substantially improved and difficult or unsafe workings also be exploited (Thring, 1983).

The world's economically exploitable coal reserves could be greatly increased by wider adoption of other advanced technologies, especially underground gasification. This technique is particularly valuable for the exploitation of reserves in deep mines (800 to 1,000 m), and thin and fractured seams. Technical feasibility was demonstrated as far back as the 1930s in the USSR, in mines in the Donetz Basin, near Moscow and in Uzbekistan. One factor discouraging the wider use of the method is the low calorific value of resulting gas. Gas quality and quantity are both difficult to predict and to control, even over the short term. Results obtained are never strictly comparable, even between mines which possess many of the same characteristics. A negative economic factor is represented by the high cost of drilling of the many boreholes required to ensure combustion and extraction. Studies and tests are currently underway in various countries, among them the United States of America, Great Britain, France, the Federal Republic of Germany, Belgium and Poland.

Conventional mines, using some of these advanced technological improvements, are still being opened up in many areas of the world and — in particular in developing countries — the fully automated mine is still a considerable way ahead in the future.

Transport poses considerable problems. By sea, the largest bulk carriers are used for reasons of economies of scale; the berthing, loading and unloading of such ships often requires the construction of special facilities. All long-distance movement of coal is therefore costly: the f.o.b. cost of American or Australian coal in Europe can be as much as two or three times the cost of getting it out of the ground. Overland transport is mostly by railway, using both special-purpose trains and special-purpose lines; some reduction in the cost of rail transport has recently been obtained by large-scale automation of both trains and lines. The use of barges on inland waterways is another frequently adopted and convenient method.

The transport of pulverized coal by pipeline, on the other hand, while in principle being cheaper, requires large quantities of water to form a diluted slurry and raises problems of cost, of yield, and treatment of the water discharged, which still contains fine coal in suspension. There are pipelines hundreds of kilometres long currently in operation in the United States of

America and in the USSR. Nevertheless, slurry pipeline transport of coal is considered economic only where very large quantities of coal are involved. Construction of a trans-European pipeline from Poland to Trieste on the northern shores of the Adriatic in Italy, for example, is under consideration. Other liquids, not sharing the problems presented by water, are currently being tested as carriers for coal. A coal pipeline remains, however, a very major commitment, and more intensive use of any existing rail network would generally prove more economic.

The impurities in coal, only susceptible to a limited extent to removal and treatment after extraction, present a drawback which coal combustion technologies are now tackling seriously. To improve the conventional combustion process, the temperature within boilers must be kept as uniform as possible. The best results are achieved by fluidized bed combustion, which moreover retains some of the heavily polluting sulphur content of the coal in the bed by the use of suitable additives. As a result of stricter standards for the discharge of gaseous waste, techniques for the removal of sulphur dioxide from boiler flue gas are the object of intensive research; the most modern plants which have been developed can reach efficiencies of 95% and permit recovery of the sulphur. Research is also being carried out on processes for the full purification of flue gases, so as to reduce substantially emissions of sulphur, nitrogen oxides, soot particles and dust.

An alternative technology permitting considerable reduction in environmental impact during combustion consists of the mixing of finely ground coal with a liquid — such as water or oil — to form a thick slurry. In Sweden, a 250,000 ton per year plant has been in operation since late 1984, after pilot trials lasting three years. Boilers formerly fired with oil and natural gas need to be fitted with specially adapted burners to take this liquid coal, but these are easily applied to most existing systems (*Coal Slurry Combustion and Technology*, 1984).

However efficient the combustion process, use of coal in solid or thick slurry form must be limited to firing blastfurnaces in steelmaking after reduction to coke, or in large boilers for electricity generation, cement kilns, district heating systems and other large-scale applications. Increased market penetration for coal is linked to ways of turning it into a gas or into a free-flowing liquid, as it could then become truly competitive with natural gas and oil, in terms of cost and convenience, since the infrastructures already in place for both transport and distribution could be utilized. Coal conversion processes are therefore the object of considerable attention. Plants for the conversion of coal to a liquid 'syncrude' exist, several of these — in the Republic of South Africa, for example — have been operating on an industrial scale for some time, financed for strategic reasons. Coal liquefaction can be achieved by different processes: synthesis, degradation or hydroliquefaction.

In the 'synthesis' processes, coal is transformed essentially into carbon monoxide and hydrogen, which in turn are combined to form liquid products. In 'degradation' processes, coal molecules are partially broken up to obtain liquid products more directly. In 'hydroliquefaction', coal is hydrogenated with elemental hydrogen or a donor solvent, and liquid products are then extracted in an organic solvent. Only one method, Fischer-Tropsch synthesis, now remains in large-scale use, being adopted in the South African Sasol II and III complexes (Probstein and Hicks, 1982).

Techniques for the conversion of coal to a low BTU gas have been known for a long time, but applications for such a gas are limited. More general application would require a medium BTU gas, or one with a BTU value similar to natural gas (synthetic natural gas). The production of such types of gas from coal is, however, a highly endothermic process and the yield of gas obtained from the coal used is low, the rest having to be burned completely to carbon dioxide; it is not at present a very satisfactory economic proposition. One interesting development is the Cool Water coal gasification programme of Texaco Inc., which produces a medium BTU synthesis gas and is a great improvement over the conventional Fischer-Tropsch process as regards pollution control and reduction of emissions. Plants using the process employ the syngas on site to produce heat, and hence electricity; others produce chemicals such as acetic anhydride for photographic film.

As oil and natural gas prices have levelled out over recent years, interest in coal conversion technologies has decreased — they are likely to find wide application at their present state of development only in the event of a new oil shock. Coal technology is therefore at a crossroads. Massive investment is required for development of both conventional and novel uses of the fuel: the infrastructures involved in creating a new coal chain are enormously costly and complex, and take time to realize. Demand, however, is volatile, subject to potential external destabilization resulting from movements in the price of its chief rival, oil. Who will assume the risk, both financial and strategic, for the opening of mines, the establishment of transportation systems to adequate ports and harbours in producer countries, for ocean transport systems or railways and pipelines — the producer, or the consumer who may be half a world away?

Furthermore, major investment programmes in conventional solid-coal technologies already menaced by decreasing public acceptance on ecological grounds could some day be threatened by the emergence of a competitively priced new technology, be it coal liquefaction or gasification at the mine-head. This suggests a certain prudence in decision-making. In addition, if any such major commitments were to be made on any large scale their very existence and the vested interests thus created would constitute a barrier to the commercial development of a new technology. It would be deadlock:

significant departures in new technology may appear prohibitively expensive, but a major programme of investment in revamped conventional uses could prove equally problematic.

NUCLEAR POWER

Nuclear fission reactors

The use of nuclear fission to produce energy is a relatively recent development in energy technology. The principle of nuclear fission was discovered in the years 1935-40, the first experimental nuclear reactor was built by Enrico Fermi in 1942, and the first commercial nuclear reactors for electric power appeared in the late 1950s. A large number of nuclear reactors for electricity generation were ordered by many countries between 1970 and 1975, some of which also produce industrial steam bled from the turbine.

The development of fission reactor technology has required a great deal of basic research in new fields of physics, chemistry, engineering and biology, with major scientific and technological advances achieved across the board in energy technologies, materials science, control systems and information technologies, to mention but a few. Nuclear technologies have stimulated inquiry in so many other sectors that they represent an important technological step toward for any country engaged in this area, with implications beyond the energy sector.

Initially, the development of nuclear energy explored many combinations of reactors and fuel cycles. Later there was a convergence on the uranium-plutonium cycle, on water-cooled thermal reactors, and on sodium-cooled fast breeder reactors. Thermal reactors now contribute between 10% and 60% of the electricity supply in countries actively following the nuclear option. World production of electricity from nuclear power doubled between 1977 and 1983, and seems destined to grow at a rate of close on 10% a year for the near future

Table 8. Production of nuclear electricity.
(billion Kwh)

	Western industrialized countries	Centrally planned economies	Developing countries	World total
1977	464.5 (90.3%)	45 (8.75%)	5 (0.95%)	514.5 (100%)
1980	601.0 (84.4%)	94 (13.2%)	17 (2.4%)	712.0 (100%)
1983	853.5 (83.9%)	132 (13.0%)	31 (3.05%)	1016.5 (100%)
1984	1006.5 (83.9%)	142 (11.85%)	51 (4.25%)	1199.5 (100%)

Source: ENEA (Rome).

Table 9. Percentage share of nuclear source in electricity production.

	Western industrialized countries	Centrally planned economies	Developing countries	World
1977	9.8	2.65	0.75	7.2
1980	11.5	4.75	2.35	8.8
1983	15.9	6.25	3.00	11.95
1984	17.9	6.55	4.6	13.45

Source: ENEA (Rome).

(Tables 8-9). The contribution of nuclear electricity to world energy consumption rose from 0.8% to 3.4% between 1973 and 1983, making an important contribution to lessening the share held by oil.

The size of the reactors built has progressively increased, for reasons of economies of scale, from the initial 200 Mwe to the present 900-1,300 Mwe. The capital cost per unit of power and the construction times for nuclear reactors are much greater than those for thermoelectric power stations. Fuel costs, on the other hand, can be less. There are certain economic advantages for a large grid system to cover base load with power generated in nuclear reactors, where these latter manage to optimize efficiencies. Given this condition, it is perhaps inevitable that very different economic performance statistics for nuclear power stations have been obtained recently, between one country and another, and even within the same country (Brookes and Motamen, 1984).

The poor economic performance of some nuclear power stations, in the United States of America and in certain developing countries, overcapacity in electricity generation owing to past overestimation of demand, and concern over environmental and safety issues are some of the factors which have caused a cutback in orders for new plants and stricter control over the operation of existing ones in many countries in the industrialized West. The nuclear industry, which already suffered from confusion in the public mind with atomic weaponry, lost more credibility with incidents tarnishing the remarkable safety record that it had built up through rigorous testing and quality controls. This occurred despite the fact that these incidents — although arousing enormous publicity — did not imperil human life or actually cause significant damage to the environment. The nuclear industry's safety record therefore objectively remains the envy of most conventional technologies.

In the near future, with the completion of power stations now under construction and some new orders, the share of nuclear power in base-load generation could start to approach optimal levels (Evans and Hope, 1984). A major obstacle to this is provided, however, by the escalation of capital costs for plant construction — in part owing to historically high real interest rates, in part to cost overruns and provision for increased safety margins — which has put back considerably nuclear's achievement of commercial profitability (Weinberg et al., 1984). The use of nuclear to follow the variation of load on the grid has recently been tried out in France, but an overall evaluation of this will not be possible for some time. The share of overall world energy consumption covered by thermal nuclear reactors could rise to about 6% between now and the year 2000.

The introduction of nuclear reactors into developing countries is hindered by the limited size of their electric grids, by lack of sufficient capital for investment and by the difficult balance between the enormous effort a society must make to assimilate nuclear technologies and the leap in quality that society makes in doing so. The development of cost-efficient smaller, modular reactors could encourage the introduction of nuclear power into the developing world to generate electricity for the larger urban areas. Until that time, however, the nuclear option is not likely to be taken up by most developing countries. Still, several of the larger developing nations have begun to explore the use of nuclear technologies for electricity generation and have plants in operation or under construction, such as India, Pakistan, China, Argentina and Brazil.

In the longer term, further penetration of thermal reactors into the world energy system will depend not only on economic, political and social factors — above all, on the level of public acceptance — but also on the solution of technological problems relating to the fuel cycle and the temporary storage and long-term disposal of radioactive wastes. Certain of the developments in this direction are connected with the introduction of sodium-cooled fast breeder reactors. Fast breeder reactors allow the amount of energy produced from a given quantity of uranium to be multiplied by a factor of 50 or 60, giving a further cushion to mankind against the day when fossil fuels are exhausted.

Fast breeder reactors may eventually supply a large proportion of man's energy needs, the energy produced being distributed for final use as electricity or as hydrogen. Sodium-cooled fast breeder reactor technology, however, has so far been tested in only a few industrial-scale plants, of less than optimal size. At their present stage of development, they are not economically competitive either with thermal reactors or with coal for electricity generation. Continuing development of these reactors is in any case motivated by strategic considerations of energy self-sufficiency for countries lacking adequate fossil fuel reserves, and is regarded as a long-term approach to the energy problem. It calls for extensive international cooperation, such as that decided in January 1984 by six European governments on the joint development of fast breeder technology.

Nuclear fusion reactors

The fusion reactor could prove a turning-point for mankind, with the production of energy from raw materials like water and lithium ores which are available in large quantities at negligible cost. Fusion would represent a real possibility of liberation from exhaustible energy sources. If commercial fast breeder reactors are still far off on the long-term energy horizon, fusion belongs almost to the realms of futurology. We are talking not in terms of years, but of many decades before its full potential can be harnessed safely by man.

The construction of nuclear fusion reactors involves tremendous difficulties. It is, in fact, necessary to reproduce in large industrial plants the conditions, in terms of temperature and material density, that occur under the surface of the sun, where at about 100 million °C nuclei of light atoms fuse together, liberating enormous amounts of energy. The creation of such an environment is a highly challenging task. It is even more challenging to keep fusion reactions under control, thus converting the heat produced into steam and then into electricity. Given that, as we have seen, the cost of raw materials is virtually nil, energy production is the result of a complex technology making use of special materials and extremely powerful magnets, based on highly sophisticated scientific concepts especially in the new field of plasma physics. Science and technology, in this case, can truly be said to be creating resources *ex novo* — a gigantic step indeed.

The development of nuclear fusion reactor technology is thus a very long-term commitment — perhaps it will be possible to produce the first commercial power in fifty years. Many advanced industrial economies are involved, either singly or as part of groupings such as the European Economic Community, because of the extraordinary challenges fusion research presents, and because of the responsibility they bear towards future generations in need of new non-exhaustible energy sources.

RENEWABLE ENERGY SOURCES

The technology of renewable energy sources includes sectors in different stages of development and of application.

Hydroelectric and geothermal power

Hydroelectric power has been exploited for many decades in developed countries and its potential there is practically exhausted (Deudney, 1981). At the world level, there are still possibilities for constructing large hydroelectric schemes (particularly in Africa, South-East Asia and South America). Likely areas are generally sparsely inhabited, often heavily forested and far from any centre of development with substantial demand for electricity.

The construction of highly automated mini-hydroelectric plants may play a certain role in the future, especially in inaccessible parts of the Third World. These exploit small falls of water and can make small isolated localities self-sufficient in their electricity supply. Full exploitation of the hydroelectric potential of the major river systems, however, would require a new international distribution of industrial production, with the transfer of energy-intensive industries to these areas of high hydroelectric energy availability, or else progress in techniques of long-distance electrical power transmission. The latter would make it economic to connect isolated centres of power production to distant markets for energy. Considerable technical difficulties exist, not least with the cryogenic systems now under examination, as well as complicated problems of international law. There is also an absolute need to identify an enforceable means of guaranteeing continuity of supply across frontiers — a vital element given that electricity cannot be stored and any interruption would have immediate effect. The maximum international cooperation is called for. International trade in electricity, which was once a purely domestic product, has been developing in recent years, both in the form of joint enterprises between countries for the construction of nuclear power stations or large hydroelectric schemes and an exchange of electric power between national grids to maximize load-factors at peak times.

Geothermal power has also been used for electricity generation for decades. In its more conventional forms — steam or hot water which, under pressure, is forced out of fissures in the Earth's surface — geothermal energy is a rather modest source of energy production. The exploitation of deep, dry, hot rocks by introducing water into natural or induced fractures is a potentially significant option but it entails some development of new technologies. It could make a greater contribution if the search for suitable geological structures is intensified and if indications of likely heat reserves in the results of prospecting and drilling carried out for fossil fuels are made public. A recently developed technology is the exploitation of underground hot water to heat dwellings and farm buildings and greenhouses. Heat pump technology can expand applications of this source relatively easily.

Other sources

New technologies of other renewable sources (solar, wind and biomass) have hitherto made a marginal contribution to the solution of energy problems but have a great potential in the longer term (Szokolay, 1984). The energy crisis provided the impetus to study the possibilities offered by the development of these sources. Many possible applications have now been studied, which often lacked much previous evaluation of their economic or social framework. Today, a critical review of possible applications for renewable source technologies is under way in many countries, to identify sectors on which efforts should be concentrated and technologies to be developed in order to increase the efficiency, reliability, profitability and large-scale applicability of plants.

Exploitation of renewable energy sources in industrialized countries is sometimes more problematic, given characteristics which can appear to conflict with established social behaviour patterns. Renewable energy is energy which generally appears in a widely diffused and dilute form. Production of such energy requires large land areas and this competes with other forms of land use in densely populated countries. In several developing countries, its use involves a multitude of small plants designed to meet particular user requirements. Their selection, construction and operation generally requires an extensive involvement of the family or of a small community (for example, the residents of a block of flats). This is not easy to obtain from people accustomed to the convenience of the energy systems in developed countries, which generally require only the paying of the bill.

Renewable energy source technology is often easier to apply in specially designed infrastructures, rather than through costly and complex adaptation of existing ones. It therefore penetrates slowly where existing infrastructures are already adequate or are of recent construction and have long renewal times, as, for example, the housing stock in many West European countries. On the other hand, the systematic application of various advanced renewable source technologies may allow small communities, farms and even factories to be more or less self-sufficient in energy.

The application of renewable energy source technology is easier in developing countries, where there are large rural areas with isolated communities, often lacking in basic amenities, and where the social structure regards their introduction more favourably. These technologies must facilitate applications suited to basic energy needs, with low-cost plants that are easy to instal and operate, and which do not require specialized personnel for maintenance (Twidell, 1981).

These overriding considerations lead to a closer examination of the uses of solar power, and then wind power and biomass. Technologies for the use of *solar energy* at low temperatures (for sanitary hot water, room heating, crop drying) has found wide application. Often, economic incentives have been provided to encourage installation of plant, so as to overcome a persistent cost disadvantage over the short and medium term for the individual user. This disadvantage is keenest where a distribution system for natural gas is already in place. Efficiencies and operational durability of solar collectors remain points at issue, and the bewildering array of different types and models offered by the large number of manufacturers can confuse the uncertain.

Technologies for the use of solar energy at medium and high temperatures have proved too costly if applied to the production of electrical power in thermodynamic cycles. Research is continuing — the European Economic Community and the United States of America, for example, are both active in this field — but major problems have yet to be resolved. In 1983, there were sixteen pilot plants for the production of electricity from solar radiation in operation worldwide, in developing countries (Mali, Kuwait, Mexico) and in advanced countries (Australia, France, Italy, Japan, the USSR, Spain and the United States of America). Their net output of electricity ranged from 27 to 10,000 kw. Design net efficiencies were in a range from 2 to 20% (Szokolay, 1984). Better prospects, however, seem to exist for application in the production of process heat.

The most promising method for the production of electricity from sunlight at present appears to be solar photovoltaics. Solar photovoltaics developed when solid state physics explained the behaviour of electrons and of impurities in solids, and the interactions between matter and light. Like the advanced nuclear technologies, the production of energy derives from the technology itself, since the sun is a free resource. Microwave relay of solar energy captured by photovoltaic cells situated on a space power plant in a geostationary orbit above the Earth remains perhaps the most futuristic aspect of the solar option. However, the application today of photovoltaic technology to resolve the energy needs of the developing world, and village life in particular, is becoming one of our best prospects for the blending of advanced and traditional technologies, a key component in acceptability. Requiring minimal maintenance, once installed providing an uncomplicated, easy to explain and trouble-free energy source, solar photovoltaics provide a valid alternative to diesel generators — and above all do not necessitate any call upon the cash economy, a vitally important consideration in the poorest subsistence areas.

Advantages which photovoltaics share with other technologies for the exploitation of renewable energy sources include low environmental impact (though large-scale uses can have an effect on the micro-climate and do take up space), availability of the primary source (solar radiation), and abundance of raw materials to construct the photocells. Specific advantages include great potential for significant cost reductions through the use of new materials (polycrystalline or amorphous silicon, or intermetallic compounds) and new production processes to obtain higher production volumes, using, for instance, microelectronics and automation. Learning curve benefits are waiting to be reaped in this field. A further advantage — above all for decentralized applications at the village level — is its intrinsically modular nature, being thus able to satisfy different levels of demand from fractions of a watt upwards. This characteristic has permitted the development of a strongly expanding market for photocells, rising from 100 kw in 1960 to about 18 Mw in 1984. Part of this demand stems from certain low-power applications for which cost is not a determining factor, such as calculators and gadgets and toys. Cost reductions have already made them economically competitive for applications in many isolated localities and unmanned sites such as road signs, ocean buoys, communication masts, where the other sources' transport costs weigh heavily.

Wind energy has made great progress in recent years. High technology windmills for electricity generation of varying outputs have been tested. These incorporate the latest computer-aided design techniques and new materials to reduce wind resistance and ensure lubrication-free operation over long periods of time. Sites are selected only after exhaustive computer analysis of meteorological data on wind speeds, directions and constancy. Quite large batteries of wind-electric generators are already installed and able to produce considerable quantities of electricity. Fortuitously perhaps, countries in the temperate and higher latitudes, which lack possibilities for the exploitation of solar power on a large scale, are those best endowed with wind potential. For them, wind power is the most likely of all the renewable sources to be economically competitive with fossil fuels for electricity generation. In many developing countries, effort has also been dedicated to the development and spread of mechanical windmills for agricultural uses. In the design of such windmills, specific attention has to be paid to the availability of materials, the ease of construction, operation and maintenance, and to the matching of local power demand profiles and wind patterns (Icerman and Morgan, 1984).

Technologies for the production of energy from *biomass* have made great progress. The contribution being made in one Third World country, Brazil, by the use of sugar cane to produce a liquid fuel (gasohol) for mass transportation will be described below (pp. 00-00). Biomass can have a variety of origins. It can be produced in farming as residues from crops intended for human or animal consumption (straw, pruning residues, rice, husks, etc.) or as plants — both land-grown and aquatic - expressly cultivated for energy purposes, or energy and food purposes combined (sacchariferous plants, plants rich in starch, etc.). Forestry is also a major contributor. Wastes of various kinds — sewage of animal and human origin, urban refuse — are other important feedstocks. Depending on its nature, biomass can be burned directly to obtain energy as heat, or else be converted into fuels by means of micro-organisms (fermentation, bacteria such as yeast, for example), or the joint action of high temperatures and chemical agents. Such conversions give a wide range of products: methane, low and medium BTU gaseous mixtures, ethanol, methanol, oils and carbon-content products. All these have wide uses. The alcohols — ethanol and methanol especially — can be blended with gasoline as octane-boosters, which saves oil products and eliminates the need for the addition of tetraethyl lead (a major environmental pollutant) as an anti-knocking agent. Another environmental benefit comes from the use of agro-industrial effluents (sewage, distillery wastes, etc.) to generate biogas via anaerobic digestion.

Advances have been made especially in the design of multi-use boilers that are able to burn agricultural and urban wastes, or mixtures of the two and conventional fuels, with greatly reduced environmental impacts. Further progress needs to be made in filters and purifiers, in particular to ensure that systems burning extremely malodorous town refuse — often themselves situated in or near urban areas — retain both the smells and the toxic wastes associated with such refuse. These latter include the heavy metals and even toxins such as dioxin and various poisonous gases liberated in the combustion process. Smoke stack technology itself is undergoing a period of renewal, and these problems should be nearer solution.

Biogas plants are benefiting from the rash of innovation which is occurring worldwide. Efficiencies can be increased dramatically by the adoption of new materials for seals and coatings, by applying simple microelectronic control devices and by techniques such as pre-heating, mechanical separation and mixing. The resulting digesters and generators are still easy to operate, with the widest possible applications. Again, this is an example of the blending of advanced technologies with those traditionally available so as to maximize returns. At the level of a small community or indeed individual farm, digesters can satisfy the bulk of energy requirements if managed properly. The aid of personal computers in management has already been applied in a large farm complex in Italy, with extremely positive results.

Biomass technologies achieve two goals simultaneously: they dispose of waste products and produce energy. In addition, they reduce pollution and make available generally highly active fertilizer derived from process residues. This last benefit can even be derived from plants burning exclusively urban refuse, after suitable sifting and separation procedures. In many cases it is possible to produce heat, in the form of hot water or hot air, for a variety of heating and drying purposes. Gasifiers for the production of mechanical or electric power with a capacity of 10-100 kw are in the industrial development stage, and can be applied with particular advantage in rural areas of developing countries. Chipped wood burners, for example, are already extensively used in North America and Brazil. It is also possible to link different renewable source technologies, for example solar systems with biogas systems, making the latter independent of an electric grid.

Irregularities in the availability of solar and wind power (cloudy days, calm days), and their unpredictability, have necessarily led to the parallel development of energy storage technologies. Alongside traditional technologies of electrical batteries and insulated mass heat stores, solar ponds can be singled out for their technological novelty.

ENERGY CONSERVATION

Energy conservation has already made a notable contribution to facing the problems raised by the energy crisis, especially in developed countries. The most effective energy conservation measures, when judged in terms of percentage reduction in energy consumption per unit of value added or consumed, have been those taken in the industrial sector.

Increases in the price of energy in many cases led to substitution of increased capital and labour for part of the energy input hitherto required. This has been achieved by the installation of energy recovery equipment, often using previously known if unadopted technology, and the appointment of energy managers and specialists in energy diagnosis to look into all realistic ways of making energy savings. A decade after the first sharp increase in the price of energy, this spontaneous adjustment to new market conditions has exhausted many of its possibilities, particularly in those cases where cost-benefit analysis strongly indicated that any capital invested in conservation would be recovered in a few years.

In many developed countries, legislation has subsequently been adopted for active encouragement of energy conservation. Such legislation should result in conservation measures being taken for which the pay-back period is longer and the investment hence less profitable. The cost of any subsidies involved is deemed justifiable given social benefits (for example, relief for the balance of payments) or strategic considerations (reduced dependence on energy supplies from politically unstable areas).

In the longer term, further energy saving could be obtained in industry through introduction of new processes consuming less energy and new products with lower energy content, through greater use of technologically advanced services and information technologies. Energy conservation is thus intimately linked to the restructuring induced by technological progress in the industrial sector of developed countries, making it difficult to measure the effect of energy conservation in the strictest sense, as opposed to that of more general changes throughout industry and society.

Historically, agricultural development aimed essentially at increasing productivity per worker and per unit of cultivated land. Such increases were obtained through mechanization and the extensive use of fertilizers, herbicides and pesticides — namely agro-chemicals with a high energy content. Drawbacks of these products can include soil degradation, increased pollution, indiscriminate extermination of wildlife, disturbance of the micro-climate and risks to human health both in the preparation and handling of certain of the chemicals.

Agricultural practice must now be revised in the direction of energy conservation and greater protection of the environment. Research aims to optimize the use of agro-chemicals. Nitrogen fertilizers now used by the millions of tons may in a few decades be replaced to a large extent by the introduction of the nitrogen-fixing gene directly into the chromosomes of cereals and other useful crops or by the achievement of artificial symbiosis at the plant root of nitrogen-fixing bacteria such as *Rhizobium*, as occurs naturally in legumes. Plant genetics is developing new strains with reduced need for agro-chemicals. Another advance is the trend to integrate chemical pesticides with biological pest control techniques, so as to achieve higher specificity of action, and thus lessen the danger to man and his environment. These new technologies permit techniques requiring lower energy use to be adopted safely and efficiently (see also Chapter 3).

For domestic purposes, energy conservation can be achieved principally by changes in the design of buildings: improvements in thermal insulation; maximization of exposure to the sun, and heating systems which encourage families to save energy are the main tools available to the architect. Heat and electricity co-generation technology and district heating for buildings are already in operation in colder climates and allow significant savings of energy to be obtained. New technologies may make this form of energy-saving more generally economic.

In transport, much has already been done through the application of available technologies to increase the efficiency of engines and to reduce the air and surface resistance of vehicles. Much can still be done to encourage greater use of public transport, which consumes far less energy than most forms of private transport, both in urban areas (underground and surface systems) and over medium and long distances (renewed reliance on railway networks). Efforts also need to be made to increase utilization of navigable waterways for the movement of goods.

In the longer term, considerable energy savings in transport will be made possible by the design of new propulsion systems with higher efficiencies than those now available. The employment of sensors and microprocessors can give more accurate control of engine operation and transmission. New materials significantly reduce vehicle weight, contributing much to greater mileage. Extensive substitution of ceramics for metals in the engine itself can enable higher combustion temperatures and hence even higher engine efficiencies to be attained (MIT-International Automobile Program, 1984).

Further ahead, the high cost of energy will contribute to the evolution of a new way of living and working, with the full exploitation of all the potentialities of computer-based telecommunications and the other information technologies. Land management, town planning, architecture and the design of transport systems will necessarily make this adjustment. The use of new technologies will then permit a marked saving of energy throughout the system.

In developing countries, energy conservation often seems even more difficult to achieve, because lower initial levels of consumption can mask possibilities for reducing use still further and — given the gravity of other, seemingly more pressing, problems — attention is frequently diverted away from conservation, as it is also from protection of the environment. Nevertheless, possibilities of energy conservation are at least as great as in developed countries. Equipment for energy generation, transmission, storage and use is often old, poorly maintained and designed without any consideration of conservation requirements. Thus there are wide margins for improving efficiency in traditional energy systems without the need to adopt sophisticated technologies and without distancing practice too much from that already established over time. For instance, it has been estimated that nearly two billion people in the world use traditional methods — often open fires or rudimentary wood stoves — for heating and cooking (Hughart, 1981). Efficiencies are almost always very low, in the range of 10%, and could very easily be increased by a factor of two or three (Krishna Prasad, 1983).

Energy conservation has therefore an important contribution to make in the solving of mankind's energy problems, in developed as in developing countries. It must be incorporated from the beginning as an essential parameter in any energy system. Only then can its positive impact be maximized. Turning to conservation as an emergency measure to obtain short-term relief from problems will inevitably be ineffectual and a source of disappointment. Conservation has a primary place in the assessment of energy systems (Jakobs and Shama, 1981).

Impacts of energy technologies on health and environment

Anxieties as to possible adverse health effects on populations living near industrial and power plants have markedly increased with the coming of the nuclear technologies. The dangers of accidental exposure to ionizing radiation led to the study of potential impacts of nuclear plants on the health of people living nearby from the start of development of this field. Extensive theoretical research and experimentation have been carried out, both on the increased probability of cancer and on the production of genetic mutations. On the basis of such research, maximum levels of exposure to radiation have been set at the international level. This approach can generally be said to set out deliberately to err on the side of caution, given the peculiar nature of the risks inherent in these new technologies and the impossibility of learning from experience in facing accidents as they occur, as had happened in previous cases of the introduction of new technologies. It has since had a considerable influence in the introduction of more stringent controls and tests for many industries, including those using traditional technologies or methods of production.

The fundamental criterion in establishing maxima for radiation exposure is that the population should not be subjected to a detectable rise in the health risks to which it is habitually exposed, as a result of the operation of the nuclear plant. 'Background radiation' is present throughout the world, and is part of our natural habitat, but the maximum effort must be made to ensure that limits found acceptable as background are not exceeded as a result of the activities of man (Berg and Maillie, 1981).

Similar considerations apply to the impact on the environment taken in its broadest sense (Mitsch et al., 1981). Nowadays, the physical environment seems to be facing a series of threats. In the developing countries, particular concern is being caused by deforestation, and by the over-fishing of rivers and seas. Pollution is a major problem throughout the developed world and in areas of the developing countries where rapid urbanization and industrialization are taking place. Macroscopic effects are becoming obvious to everyone.

Acid rain is one of the most alarming phenomena of pollution. It seems to arise mainly from the inefficient application of energy technologies. Like many other forms of pollution, it ignores national frontiers and geographical boundaries: it must therefore be tackled within the framework of extensive international cooperation. This should first concentrate on recording all data required in order to establish the nature and intensity of the phenomena, and to conduct research into the mechanism of its formation. It is important to determine the role which the various applications of energy technologies play in provoking acid rain: power stations, domestic heating and road vehicles. At present, the contribution of each of these factors is unclear. Countries will need to be prepared to apply suggested remedies in a coordinated manner, and the costs will probably have to be shared between all those concerned, given that the effects of acid rain are felt in countries which may have only a very limited responsibility for its formation.

Substantial changes in the earth's climate could be threatened by the so-called 'greenhouse effect', which some environmentalists believe is being caused by a gradual accumulation of carbon dioxide in the atmosphere as a consequence of the burning of fossil fuels and of the progressive deforestation of the Earth. Its mechanisms are again really still little known, and much further research will be needed. Concern has been aroused especially by the plight of delicate forest and lake ecosystems in the temperate zones of the northern hemisphere; but actually the possible role to be assigned to the indiscriminate destruction of the equatorial forests means that both the greenhouse effect and acid rain are not problems confined to the industrialized world. The global long-term economic effects could prove considerable, and for this reason they must be tackled internationally.

As a result of the increasing gravity of environmental problems, and the sensitivity of world opinion to ecological issues, it is now recognized that we must try to develop new ecotechnologies in all our industrial activity, especially in that related to energy. These ecotechnologies give rise to an entirely new sector of industry, the development of which should be given high priority: the protection of the environment from the deleterious effects of human activities, and the identification and diffusion of systems which permit industrial production without negative environmental impacts.

Governments and international organizations are entrusted with very important responsibilities in this regard. Before choices are made, for which the usual terms of reference of cost-benefit analysis provide no support, it is important that public opinion should be kept fully informed of the possible long-term results of policy decisions, however seemingly unquantifiable they are. Moreover, it is absolutely essential that national regulations governing environmental protection be harmonized, to avoid conflict and any undue disruption of world development.

Planning for the energy needs of the developing world

Creation of the infrastructures necessary to improve the standard of living in developing countries often calls for the application of considerable amounts of energy, generally imported at considerable cost to the balance of payments. Development in itself tends to be linked to an increase in the demand for energy, since economies tend to need yet more energy to make proper use of these infrastructures, once acquired; continuing expenditure is then involved in their maintenance and servicing. Many difficulties have to be overcome in order to meet this enhanced demand. Development of domestic resources is often hindered by lack of capital, technology and infrastructure, while imports of energy are limited by balance of payments constraints.

Energy policies have to be formulated with great care if they are not themselves to become obstacles to development. They have an urgency of their own, despite the myriad intractable problems besetting these societies. Energy planning is neither simple nor easy. In many cases, the data required is simply not available — this applies particularly to data on the use of non-commercial energy sources, such as animal power, firewood, and agricultural and animal waste. Often the first step has to be the collection of reliable data, using the best methods including sample surveys and the sharing of experience in other countries (Neu and Bain, 1983; Meier, 1984).

Energy planning in developing countries cannot be carried out by following general schemes. Account must be taken of the specific characteristics of each country and the plan ought to be prepared by the people of that country. Many, perhaps the majority, of countries have overall development plans within which an energy plan has to be fitted. The energy plan must further the objectives of the general development plan, even if the working out of the energy plan may lead to a revision of the general plan itself — should this create, for example, otherwise insoluble energy supply problems through an overemphasis on energy-intensive activities. Energy planning must also not neglect the social characteristics of each country, adapting to its structure, habits and needs. Furthermore, it must keep well in mind the central role still to be allotted to agriculture in most developing countries, and their overwhelming priority to achieve self-sufficiency in food production.

PROGRESS TOWARDS SELF-SUFFICIENCY

Notwithstanding these difficulties in planning and management, some progress towards energy self-sufficiency and towards reducing the oil share has been achieved in developing countries in recent years. Much still remains to be done, however, given constantly increasing energy demands from the developing world. Finance from the World Bank and other international and regional financing bodies has helped with the exploration and exploitation of deposits of oil, coal and natural gas. Between 1973 and 1983, oil output approximately doubled in China, Brazil and in West Africa (excluding Nigeria). It tripled in Egypt and in the countries of the Indian subcontinent, and quadrupled in South-East Asia (excluding Indonesia and Brunei).

In some countries, this oil output has been used domestically, to maintain or increase the level of energy self-sufficiency even in the face of rising consumption. In others, it has made an important contribution to exports, going some way to pay for the increased importation of capital goods necessary for industrial take-off. In this latter case, a well-judged policy governing domestic prices for oil products may balance an increase in investment made possible by rising exports, with improvement in living standards — stimulated by increased availability of energy. Exploration and exploitation have at times been impeded by social and political instability. Investment on this scale needs a continuing public commitment to development goals.

During recent years, coal production in developing countries has increased markedly, most notably in Turkey, Colombia, India and China. Natural gas production has been increased significantly in Southern Asia, China, Bolivia and Brazil. Between 1973 and 1983, coal production in the developing countries rose from 8.45% to 11.7% of total world output; natural gas production from 9.35% to 14.45% and oil production in non-oil exporting countries from 5.0% to 8.45% (Tables 4, 6 and 7). In the decade after the 1973 oil crisis, hydroelectricity output approximately doubled across Latin America, Southern Asia, China and Africa. Its production is often limited, however, not by the lack of generation plants but by inadequate transmission grids, unfavourable location of potential sites and insufficient electricity demand to justify the capital outlay.

Application of new technologies for renewable energy sources in developing countries is mostly still experimental in character. However, the use of biogas in China, the Indian subcontinent and South-East Asia and the use of ethyl alcohol or a mixture of alcohol and gasoline (gasohol) in internal combustion engines in Brazil are already widespread and may have much greater potential. By 1984, over 4 million small biogas digestors were operating in China for household (81%) and agricultural (19%) purposes, with an annual output of a million tons of coal equivalent; on average, each plant produces approximately 3 kg of coal equivalent a day. The growth potential for such biogas generators is obviously still enormous. In China alone, the consumption of bio-energy in rural areas has been estimated at 200 million tons of coal equivalent per year.

In Brazil, the national alcohol programme already operates on a large scale. By 1980, all gasoline sold contained 20% alcohol — it having been established that the engines of the existing car stock could run on this mixture with only minor modifications — and over 3,500 pumps were supplying ethanol as a fuel. The programme was further strengthened in 1983, when the alcohol content of gasoline was increased to 22%. 70% of new vehicles sold in 1984 were able to run on 100% alcohol as fuel. The programme envisages that all vehicles on Brazilian roads by the year 2000 will be fuelled by alcohol. 90% of the alcohol is currently obtained from sugar cane, 10% from other crops. The expansion of the programme is made possible by dedicating more land to fuel crop cultivation, higher crop yields, the building of new distilleries and by improved techniques. The higher cost of alcohol even when compared to that of imported gasoline (about 30%) is borne by the government in recognition of the consequent alleviation of pressure on the balance of trade. Plant-derived alcohol accounted for 7.4% of Brazil's energy consumption in 1983. The energy plan envisages that this contribution will rise to 9.9% in 1993, a result that can be achieved by cropping only around 2% of the land area (Monaco, 1984).

DEVELOPMENT OF APPROPRIATE TECHNOLOGIES

Both biogas and ethyl alcohol represent important and independent developments in technologies appropriate to the geographical and social conditions in a developing country. They are both based on already established usages and procedures: the use of dung as fuel in Southern and South-East Asia; the ready availability of arable land, the cultivation of sugar cane and distillation of alcohol in Brazil. One major stimulus to the start of the Brazilian gasohol programme had been excess production of sugar cane, at a time of falling prices on world commodity markets. There are certain problems in the production of fuel by this method, not least seasonal unemployment among farm labourers, the risk of pests, the creation of an agricultural monoculture with all its negative effects on soil fertility. Whether these disadvantages will, in the longer term, have consequences for the future of the programme itself is difficult to tell at this stage, especially in the context of recent major oil finds in Brazil. In any case, gasohol and dung technologies are quite difficult to export to countries with different geographical conditions or customs.

The impact of new energy technology on the environment may be generally less severe in developing than in industrialized countries in that traditional energy sources already cause problems, such as those presented by excessive exploitation of forests and woodlands, and these effects have to be taken into urgent consideration. Deforestation and desertification are often linked in arid zones to the search for kindling and firewood. Indiscriminate burning of animal and vegetable wastes destroys valuable natural fertilizer. Energy technologies should be applied in developing countries in such a way as to minimize environmental problems present in many developed countries today.

The rational exploitation of domestic energy sources is essential for the energy development of the developing world. Of these sources, renewables (and solar energy in particular) ought to have a high priority. Many developing countries lie near the Equator, and the exploitation of diffuse form of energy, like solar energy, usually does not interfere with other claims on land use in these countries.

Energy planning for the developing world must dedicate an important share of resources to prospecting for and exploitation of fossil fuel deposits. An examination of the geographical distribution of known coal reserves shows that these are concentrated in industrialized countries and a few developing countries, not however in any proportion to their land areas. Deposits of oil, natural gas or coal could well be found, given the vast size of the developing world, but prospecting for such deposits is frequently neglected by the major multinational oil and coal companies. Finds tend not to be worth exporting, given their limited size and the scale of the investment needed to get them out of the ground and to an adequate port. Political instability, as we have seen, can undermine even the most optimistic economic analysis of return on investment. Naturally, sovereign states hope to capitalize on any resource finds, and this can complicate marketing and pricing policies for international operators. However, governments often simply do not have the financial and technological resources to undertake exploration and exploitation on a realistic scale. This is where international cooperation and assistance from richer or better endowed countries can play a part.

International cooperation needs principally to take the form of transfer of methods of analysis and planning and of energy technologies, as well as direct help with the training of technicians. The supply of production plants, often unsuited to conditions existing in developing countries, is no longer to be seen as a panacea. Possibilities for South-South cooperation should not be overlooked, and neighbouring countries can cooperate in analysing regional energy patterns leading perhaps to the joint exploitation of energy resources in border areas. Joint R&D programmes, and indeed energy development plans, should be envisaged. With this approach, it should be possible to maximize the chances of reaching the solution best suited to the characteristics of each individual country, and of making a real contribution to the development effort.

Finally, appropriate technologies are likely to be the key to the developing world's energy future, as they will probably be to development problems in general. Urbanization pressures must be contained, and this can best be achieved by making rural areas once more places where life has dignity and sense, and is not merely a desperate struggle for survival. Thus the picture at the village level matters enormously. The provision of energy at minimal cost and minimum disturbance to the local populace, their societies and often subsistence economies, is an absolute priority if the drift to the cities is to be halted. Electricity, preferably generated on site, perhaps by photovoltaic conversion of solar energy, seems to offer some hope of this — a reliable source of energy for refrigeration of essential foods and medicines, for communications and education, and even for the drawing of water.

The example of the diesel generator — which now can be seen as a costly error that has done much to disturb the pattern of rural life, and whose unreliability creates almost as many problems as are solved — should act as a warning. Appropriate technology need not be the most simple, so long as any maintenance is trouble-free and repeated calls upon the market economy are avoided. By adopting a system of this nature, vastly expensive rural electrification schemes involving grid networks can be postponed until a financially more auspicious moment. The sprawling urban areas of the developing world already need an energy supply similar to that of the industrialized world. Many developing countries therefore already have to confront energy problems as complex, if not more so, as those of the advanced economies. They require the same sophistication of analysis, the same ability to weigh up costs and benefits, and above all an even greater capacity to reduce dependence on imported sources whose pricing system is beyond their control.

Energy technologies as a factor for change

The energy system is only a part of a multiplicity of subsystems — such as the ecosystem, the social system, the economic system, and the technological system — that make up the global picture. Within each system grouping there are geographical subdivisions, according to countries and regions. All the subsystems are mutually interacting, and thus interdependent; at different times and in different countries the various subsystems can play a variety of roles in determining global change. At present in industrialized countries, the computer-based or computer-related technologies have the most widespread effects in determining change in all other technological subsystems and a short time after in the economy and society. The developments in microelectronics and their pervasive applications in fact permit far greater efficiency in the use of energy, although it goes without saying that these societies now depend upon reliable supplies of energy for every aspect of life. In most of the developing world, on the other hand, energy is still the critical factor determining the development of the other subsystems. Improvements in agriculture and industrialization – a development stage that cannot be leapfrogged — both involve large increases in energy consumption. For developing countries, apart from a handful of oil-producing nations, energy supplies and technologies must be carefully planned and developed if energy is not to be the overriding constraint to societal change.

The present energy situation provides a wide variety of technical options. At one end of the spectrum are the energy macrosystems: complex, very rigid, centralized systems involving vast power stations, extended distribution networks, an enormous international trade in fossil sources (and perhaps soon also in electric power), the building of energy infrastructures that lock participating countries into restrictive internal and external relationships. They can themselves easily become a source of rigidity in the world energy system. Macrosystems tend to require large capital outlay, lengthy construction times, and then efficient management and control systems to maintain the whole cumbersome system in operation. The consequences of breakdown or malfunction are far-reaching, and irreparable damage can be done to an economy if such a system should fail. The new information technologies provide the most relaible solutions to these problems, through the use of automated control systems and other advanced techniques which themselves have to be rigorously maintained throughout the life of the macrosystem. The scale of these systems is so vast that they have to be managed by large private or state corporations, with inevitable government involvement in planning and construction of schemes and in coping with their repercussions on society. One aspect in particular, the siting of large energy installations, may elicit protests from local populations, and this is as true of a small hydroelectric scheme in a rural area of Africa as it is of a very large nuclear power plant in Europe or North America.

There is currently no alternative to macrosystems in industrialized countries and in heavily urbanized areas of developing countries, but it is now also possible to design local integrated energy systems to meet the needs of small communities; independent and self-sufficient minisystems using all available technologies and energy resources, such as the sun, the wind, biomass or small waterfalls (Gilmer, 1981; Nader and Milleron, 1979). Minisystems possess certain features that make them inappropriate for meeting high levels of demand in densely populated areas (they tend to require large amounts of labour and or space, and economies of scale work against them), but they are well suited for use in isolated communities in both the developed and the developing world that lie beyond the feasible reach of the macrosystems' distribution networks. In particular, appropriate minisystems can provide at low capital cost quite adequate energy supplies for agricultural communities and for the early stages of development, and hence they can play an important part in halting the drift of population from the countryside to overcrowded urban centres (Twidell, 1981).

Later stages of development may require more sophisticated energy sources, and macrosystems supplemented by minisystems can together contribute to the stability and flexibility of an energy system as a whole.

It is not a simple task to select the best solution from all the varied possibilities offered by energy technologies, against a background of uncertainty over the future development of technology and trends in energy prices and demand. Estimating technological and economic trends leads at most to the identification of areas of possibility and probability distributions, varying from country to country, often from one part of a country to another. It is therefore advisable to design energy systems incorporating as high a degree of flexibility as possible and to avoid excessively one-sided solutions.

Developments in energy technologies are backed by parallel advances in the scientific, economic, organizational and socio-political fields. Solid state physics, materials science, the theory of automatic controls, biotechnology, physical chemistry and plasma physics are among the branches of science which now play important roles in energy development. This aspect of energy policy is one which is of particular relevance and responsibility for the developed world. Many scientists from developing nations have doubted the wisdom for their countries of getting involved in the development and application of ultra-high tech solutions to energy problems. They see more scope for the refinement of traditional technologies, their blending with new technologies to produce a mix adapted to the needs of particular societies, and concentration of research efforts on areas of greatest hope for the developing world.

The planning and development of the energy system in developing countries provides an opportunity for a leap forward in the quality of life, and also in the ability to tackle the complex problems besetting modern society — indeed the systematic approach to energy problems, taking into account the economic, social and natural environment, itself provides a good training for the solution of more complex issues. The experience of developed countries is there to be drawn on but, as with other technologies, it must be used with caution, given differences in the basic structure of society. In recent years, it has become clear that developing countries cannot delegate the planning of their own energy development to individual experts or teams of experts from developed countries, rather they must assume the responsibility themselves. This requires mastery of advanced technologies, research methods and methodologies, and the ability to coordinate these tools and adapt them to local society. Even in developed countries these skills tend to be found only in large organizations, whether governmental or private sector. New forms of international cooperation, by which such skills can be shared with developing cuntries, are needed.

Developing countries should try to give energy problems a sufficiently high priority to ensure the realization of an energy base which is without doubt indispensable for development. Instruments available to a government to stimulate energy development are varied, and range from direct regulation, to tariff policy and economic incentives. Above all, it requires coordination and long-term policy-making — neither easy to achieve. Governments and international organizations can encourage research, provide finance and training, and further international cooperation. As the development of energy systems generally requires vast amounts of capital, the world financial system has a key part to play in funding.

In reviewing the new energy technologies, and the changes which they lay encourage or promote in society, the need for greater international cooperation emerges clearly. Many energy problems, just like so many of the other issues connected with the development of a society, have a world dimension. They can best be solved in a spirit of partnership and friendship between all nations.

References

BERG, G.G.; MAILLIE, H.D. (eds.). 1981. *Measurement of Risks*. New York/London, Plenum Press.

BROOKES, L.; MOTAMEN, H. 1984. *The Economics of Nuclear Energy*. London/New York, Chapman and Hall.

CHATEAU, B. 1985. La prévision énergétique en mutation? *Revue de l'énergie* (Paris), No. 370, pp. 1-11.

Coal Slurry Combustion and Technology. 1984. *Proceedings of the Sixth International Symposium*. Pittsburgh, United States Department of Energy, Pittsburg Energy Technology Center.

COLOMBO, U. 1982. Alternative Energy Futures: The Case for Electricity. *Science* (Washington, D.C.), No. 217 (20 August), pp. 705-9.

COLOMBO, U.; BERNARDINI, O. 1979. *A Low Energy Growth 2030 Scenario and the Prospects for Western Europe*. Report for the Commission of the European Economic Community. Brussels.

DARMSTADTER, J.; DUNKERLEY, J.; ALTERMAN, J. 1977. *How Industrial Societies Use Energy: A Comparative Analysis*. Baltimore, John Hopkins University Press.

DAVIS, J.D. 1984. *Blue Gold: The Political Economy of Natural Gas*. London, George Allen and Unwin.

DESPRAIRIES, P.C.; BOY DE LA TOUR, V.; LACOUR, J.J. 1984. La mobilisation progressive des ressources pétrolières, facteur de hausse modérée des prix. *Revue de l'énergie* (Paris), No. 367, pp. 627-42.

DEUDNEY, D. 1981. Hydropower: An Old Technology for a New Era. *Environment* (New York), Vol. 23, No. 7, pp. 16-21.

EVANS, N.; HOPE, C. 1984. *Nuclear Power: Futures, Costs and Benefits*. Cambridge, Cambridge University Press.

FREEMAN, C.; JAHODA, M. (eds.). 1978. *World Futures: The Great Debate*. Science Policy Research Unit, Brighton, University of Sussex.

GASS, J.J. (ed.). 1984. *Validation and Assessment of Energy Models*. National Bureau of Standards, Washington, D.C.

GILMER, R.W. 1981. Power to the People: The Promise of Decentralized Solar. *Technology Review* (Cambridge, Mass.), Vol. 23, No. 8, pp. 36-42

GRAINGER, L.G.; GIBSON, J. 1983. *Coal Utilization: Technology, Economics and Policy*. London, Graham and Trotman.

HOBSON, G.D. (ed.). 1983. *Modern Petroleum Technology*. 5th revised edition. Chichester, John Wiley and Sons.

HUGHART, D. 1981. *Prospect for Traditional and Non-conventional Energy Sources in Developing Countries*. World Bank Paper No. 346.

HYDE, M. 1981. *Energy: The New Look*. New York, McGraw Hill.

ICERMAN, L.; MORGAN, R.P. 1984. Renewable Energy Technology for International Development. *Energy* (New York), Vol. 9, No. 7, pp. 545-54.

JAKOBS, K.; SHAMA, A. 1981. Defining Energy Conservation: A Behavioural Perspective. *Energy Communications* (New York), Vol. 7, No. 6, pp. 581-2.

KEEPIN, B. 1984. A Technical Appraisal of the IIASA Energy Scenarios. *Policy Sciences* (Amsterdam), Vol. 17, No. 3, pp. 199-275.

KEMP, M.C.; LONG, N.N. (eds.). 1984. *Essay in the Economics of Exhaustible Resources*. Amsterdam, North Holland.

KRISHNA PRASAD, K. 1983. Wood Burning Stoves: Their Technology, *Economics and Development*. ILO, Geneva.

LANGLEY, K.F. 1983. *Fuel Cells. An Appraisal of Current Research and Future Prospects*. Energy Technology Research Unit, Report 12. Harwell, England.

LAWRENCE, P. 1984. Mixing Off-shore Oil and Politics. *Technology Review* (Cambridge, Mass.), Vol. 87, No. 6, pp. 38-48.

LEVY, M. 1981. Socio-cultural and Socio-economic Constraints. Paper presented at Unesco Conference on 'Non-technical Obstacles to the Use of New Energies in Developing Countries'. Bellagio, Italy.

MARCHETTI, C.; NAKICENOVIC, N. 1979. *The Dynamics of Energy Systems and the Logistic Substitution Model.* IIASA Report 79-13. Laxenburg, Austria.
MARLEY, R.C. 1984. Trends in Industrial Use of Energy. *Science* (Washington, D.C.), Vol. 226 (December 14), pp. 1277-89.
MARSHALL, E. 1984. The Procrastinator's Power Source. *Science* (Washington D.C.), Vol. 224 (April 20), pp. 268-70.
MEIER, P. 1984. *Energy Systems Analysis for Developing Countries.* Berlin, Springer Verlag.
MERRICK, D. 1983-4. *Energy: Present and Future Options.* 2 vols. Chichester, John Wiley and Sons.
–. 1984. *Coal Combustion and Conversion Technology.* London, Macmillan.
MIT-International Automobile Program, 1984. *The Future of the Automobile.* A report of MIT's International Automobile Program. Cambridge, Mass., MIT Press.
MITSCH, W.J.; BOSSERMAN, R.W.; KLOPATCH, J.M. 1981. *Energy and Ecological Modelling.* Amsterdam, Elsevier Scientific Publishing Co.
MONACO, L.C. 1984. *Integration of Emerging and Traditional Technologies in Alcohol Production.* United Nations Paper TI-32. New York.
NADER, L.; MILLERON, N. 1979. Dimension of the 'People Problem' in Energy Research and the Factual Basis of Dispersed Energy Futures. *Energy* (New York), Vol. 4, No. 5, pp. 953-62.
NEIMAN, M.; BURT, B.J. (eds.). 1983. *The Social Constraints on Energy Policy Implementation.* Lexington, Lexington Books.
NEU, H.; BAIN, D. (eds.). 1983. *National Planning and Management in Developing Countries.* Dordrecht, D. Reidel Publishing Co.
PETROSKI, H.H. 1984. Off-shore Engineering: Oil from Troubled Waters. *Technology Review* (Cambridge, Mass.), Vol. 87, No. 5, pp. 53-76.
PLUMMER, J.L. (ed.). 1982. *Energy Vulnerability.* Cambridge, Mass., Ballinger Publishing Co.
PRIEST, J. 1984. *Energy for a Technological Society.* London, Addison Wesley.
PROBSTEIN, R.F.; HICKS, R.E. 1982. *Synthetic Fuels.* New York, McGraw Hill.
RIDER, D.K. 1981. *Energy: Hydrocarbon Fuels and Chemical Resources.* New York, Wiley.
SHILLING, H.D.; BONN, B.; KRAUSS, U. 1979. *Coal Gasification.* London, Graham and Trotman.
STERNMAN, J.D. 1983. Economic Vulnerability and the Energy Transition. *Energy Systems and Policy* (New York), Vol. 7, No. 4, pp. 259-72.
SZOKOLAY, S.V. (ed.). 1984. *1983 Solar World Congress.* Oxford, Pergamon Press.
THRING, M.W. 1983. *Robots and Telechirs.* Chichester, Ellis Horwood.
TWIDELL, J. (ed.). 1981. *Energy for Rural and Island Communities.* Oxford, Pergamon Press.
WEINBERG, A.M.; SPIEVAK, I.; BARKENBUS, J.N.; LIVINGSTON, R.S.; PHUNG, D.L. 1984. *The Second Nuclear Era.* Report of the Institute for Energy Analysis of Oak Ridge Associated Universities for the United States Department of Energy. Oak Ridge, Tennessee.

6

Technological advance and educational change

Introduction

Technological change interacts with educational change across a very broad spectrum. Each and every one of the topics examined in this book — be it development in manufacturing processes, in biotechnologies, in resource or energy management, or indeed in any area of technological advance — is capable of generating major change in the social and economic fabric of society, any society. Each also demands considerable adaptation in the education system, in respect of both what is taught and how it is taught.

The precise effect of these new technologies is almost impossible to predict. Many of them (information technology, biotechnology, for example) are still very new, and there is only limited evidence on which to base assessment of their full potential. Furthermore, they are in a state of constant and rapid evolution, with the consequence that the reasonable predictions of today risk becoming the ridiculous underestimates of tomorrow, while even greater uncertainty surrounds the social changes to which they are likely to give rise. Finally, 'social change' itself is not a unitary concept: both the kind and the extent of social change which may follow from technological growth will vary from one society to another, from industrialized to developing country, from one socio-cultural context to another. Thus there can clearly be no question of there being a single, identifiable process of educational adaptation to a single, identifiable process of socio-technological change. In the field of education, perhaps even more than in other areas, the great diversity among developing (and developed) countries means an equally great variety of response to technological advance with its different socioeconomic (and cultural) consequences. The dream of a universal journey towards a set of agreed targets would be the more convincing if all concerned were starting from the same point of departure and using the same modes of transport.

In spite of these problems, it is essential to try to identify the implications of the emerging technologies for education, so that existing systems can be modified to meet new demands.

Demands are generated, first of all, by technological development itself: if this development is to continue and society is to derive maximum benefit from it, there exists an urgent demand for educating new generations to work in and with the new technologies; the range of *specific* skills required will be discussed later.

Then there are demands particularly relevant to developing countries. Technological advance is exponential; that is to say, it builds upon existing knowledge and existing capacities for development, increasing that knowledge base by approximately a factor of two in every two years. The implication is that countries already rich in their command of the new technologies and in their resource bases for further development will stride even further ahead of those countries that lack these advantages. While educational effort alone cannot bridge this gap, investment in education and training related to new technologies is, for developing countries, essential if the North-South divide is not to be further exacerbated.

Technological advance does not only create demands for education, it also provides new methods of teaching, particularly through the use of the new information technologies. These changes in techniques could have far-reaching effects on educational systems, in both the way that they function and the way that they are perceived. Although it is far from certain that such methods will be widely adopted (after all, it is generally felt that education missed many of the opportunities offered by an earlier technological advance: television (Cerych, 1985)), the impacts are more readily identifiable than the consequences of technical change for education. The main emphasis in the discussion here will accordingly be upon these 'educational technologies', deliberately focusing on northern Europe since it is there (in France and the

United Kingdom, in particular) that there has been the greatest investment to date in the exploration of these techniques, and on the compulsory rather than vocational stages of education.

Finally, it should be noted that there is an intrinsic link between the information technologies as technologies and the whole matter of education. The processes involved in these technologies — the treatment, storage and transmission of data, and their transmutation into information — can be compared to the range of conceptual skills which the development of learning gives to a child. This is not to say that the electronic processes of information handling necessarily replicate human cognitive processes; this is an area of intensive research at present in the field of Artificial Intelligence. But what is important is the very clear functional similarity (Gwyn, 1984). While the business of schooling is not *only* about the acquisition and handling of information (the development of artistic and interpersonal skills, for example, are also of the highest importance), it is true that the ability to 'learn', in the narrow sense of acquiring information, and of developing the ability to use that information, is and has been traditionally one of the main focuses of the educational process.

Impact of information technologies on the educational process in the industrialized countries

The totality of the phenomenon labelled 'education' is complex, consisting as it does of a great many processes and contributing forces which differ in kind one from the other. To illustrate something of this complexity — and without making any claim for the comprehensiveness of the picture drawn, given the variety of education systems — it may be helpful to identify the following groups of perspectives on the total picture: each and any one of these sets of perspectives could in itself provide a useful, and indeed total, framework for analysis.

CURRICULUM

The most obvious of the ways in which the advent of the information technologies affects the curriculum is through the need to accommodate new knowledge and skill acquisition within the spectrum of what is taught; indeed, this is as true for the entire range of new technologies as it is, specifically, for the new technologies of information. When we begin to probe behind the truism, however, with a view to identifying just what this new knowledge and these new skills might be, quite unexpected complexities emerge. For it is the case that many countries, especially the industrialized countries, which have made already a considerable investment in the introduction of the information technologies into education, have done so without any clear picture of what was required even in curriculum terms ('even' because the construction of a new 'subject' area, difficult though it may be, is considerably easier than introducing a new dimension into existing pedagogical practice).

It may be helpful to consider, in the sketchiest outline only, experience from four developed countries in northern Europe.

In the United Kingdom, for example, a country with a decentralized curriculum which allows a great deal of latitude to local curriculum initiatives, there emerged even in the late sixties a new curriculum 'subject' labelled, usually, 'Computer Studies' and designed to introduce children to basic principles of computational processes and to elementary coding, and also to some applications and social implications of information technology. Such innovations have produced some very interesting courses, but it could scarcely be said that their introduction corresponded to a coordinated assessment of the employment opportunities likely to arise from the spread of the information technologies. Such coordination as has emerged — from, for example, the very valuable work of the Education Committee of the British Computer Society — has focused rather upon the features deemed intrinsic to computation and its applications, and on an assessment of what the 'computer-literate' pupil should know about the new technologies. An important later development was the emergence, in 1980, of the national microelectronics Education Programme, the policy statement for which (Fothergill, 1981) proposed developments in a number of subject areas having in common the broad characteristic of being 'employment oriented'. (Subsequent developments in the United Kingdom including the strong and spontaneous demand for access to information technologies which arose at the primary level, in the event have rather distorted this original intention.)

It remains the case, however, that even in such a country, with some twenty years experience of the introduction of information technologies into education, no clear or agreed picture exists of the supposed new skill requirements. (That these new skill requirements exist is not contested; it is their identification which is lacking). As it happens, the known shortage in the work force of persons trained to work with information processing serves to muddle the issue and legitimize almost any kind of 'course' in information technology (or plain 'computing'), provided it appears to qualify students in ways which appear likely to be appropriate.

In the Federal Republic of Germany, the position is more clear-cut. The autonomy, in educational matters, of the ten *Länder* authorities and of West Berlin counsels caution when referring to developments in the Federal

Republic at large, but there does appear to be a consensus among the responsible bodies that there is a 'subject', *Informatik,* which may be introduced at upper-secondary level on an optional basis; the agreed content of the subject is heavily oriented towards an algorithmic approach to problem solving, and to codification. Put crudely, students emerging from this course are on the way to becoming computer (programming) professionals. Thus the picture is in a sense tidier than that of the United Kingdom, and it is easy to see the policy being operated as being more purposeful. Nonetheless, questions remain. Is it wise, for example, to offer students a choice between pre-professional training in computing skills, on the one hand, and on the other, nothing? How does such a policy respond to the likely need for a widespread computer literacy among all sections of the future work force? Again, it is difficult to feel that there is here a fully thought-out analysis of needs.

In Sweden, yet another emphasis is to be found (Johansson, 1985). National policy decrees that skills related to information technologies shall indeed be inculcated, but prominent among these are not so much processing as civic skills: a great emphasis is put on the need to teach future citizens how information technology may be both used and abused within the social framework. Topics to be studied include, for example, the nature and uses of data-banks, and the issues which they raise for individual privacy and its protection. Such issues will be discussed further when considering the socio-cultural setting in which the information technology education interaction takes place, but it would be wrong to dismiss the Swedish emphasis as 'merely' a cultural matter; the point is that it stems from an analysis of the skills needed by future inhabitants of an information society. The fact that these skills are defined in terms of what might be called civic competence rather than job-related skills in no way diminishes their importance: they are a legitimate — not necessarily complete — response to the educational challenge thrown down by the new technologies.

Developments in France have followed a different pattern again. Here, the first emphasis, in point of time, was on the exploration of the potential of information technology as a pedagogic tool. The French pattern was in fact almost the mirror image of that found in the United Kingdom, since the acceptance of 'informatique' as a discipline in the school curriculum came relatively late. Indeed, we find that in his influential Report to the President of the Republic, 'L'éducation et l'informatisation de la société', Professor J.C. Simon still feels the need to argue the case for *informatique* as a part of the curriculum (Simon, 1980). By contrast, the work done in France in respect of information technology as a pedagogic tool has been far-reaching, and will be discussed below.

Clearly, characterizations as brief as these of complex national developments risk being no more than crude caricatures, but the broad distinctions drawn are nonetheless valid, and the nature of the curriculum response in the four countries remains stubbornly different. Yet all four (the choice is deliberate) are countries with highly developed economies; they are grouped in close geographical proximity and they all draw upon cultural heritages which have much in common one with the other. The fact remains that there is no one agreed definition of what new skill requirements have to be catered for in the curricula which will take even the developed, let alone the developing, countries into the next century.

This is not to say that no patterns at all are emerging. There appears to be a consensus, for example, that information technologies must be introduced into education because they are seen (though in ways not yet properly comprehended) as providing a key to economic progress. This argument is clearly evident in the pronouncements of the leaders of all political parties in France and the United Kingdom, for example; it is the familiar demand for an appropriately trained work force for the future. The FAST (Forecasting and Assessment in Science and Technology) Project of the European Communities reported in 1984 that: 'The next fifteen years will be years of perhaps unprecedented changes in the skills and qualifications required in work and in private life. Education, training and retraining are the tools for bringing about such changes, and more generally for equipping the individual to command the new information technology' (FAST, 1984: 67-8).

A preliminary examination suggests the following impacts on the curriculum. It is clear, first, that there will be highly specific skill requirements created by the employment potential of all the new technologies. Given the sophistication of many of the technologies in question, the educational responses generated are likely to be met at tertiary and upper-secondary levels. What exactly these job-specific skills are going to be is a question decided very much by the pattern of economic development in any given country.

Second, however, there are two general points about future skill requirements that make possible a rather more constructive approach to planning. As suggested already, information processing is likely to be a key factor in the development of the new technologies generally, so that a thorough grounding in this area is likely to be transferable across a wide range of specialist subjects. (This range, of course, encompasses existing skill areas.) The other point is that there appears to be a consensus that, whatever the employment patterns of the future, they are likely to be characterized by greater worker mobility across sectoral boundaries than hitherto. This reinforces the belief that, as the emerging skill requirements of technologies such as, for example, biotechnology are monitored, it would be well to introduce into curricula an understanding of basic skills such as are called for in information handling: problem analysis, creation of algorithms for solution-finding, the structuring and interrogation of information, data interpretation and the communication of findings. What will be important is to avoid setting young persons at an early stage of their professional development upon narrowly-defined paths which do not allow them flexibility of career choice.

Closely allied to these questions is a third, namely the long drawn-out debate about the teaching of programming skills. It is unfortunate that the argument about programming was clouded from the beginning by the widespread use, in the developed countries, of the computer language BASIC as the point of entry. There is no argument whatsoever apart from that of expediency (because it is the language that comes with most micro-computers currently available) for teaching BASIC to children; its lack of structure leaves it with no intrinsic merit as an intellectual exercise. With other languages, however, the situation is very different. Languages such as LOGO and micro-PROLOG, which allow direct, structured operation upon an identifiable 'object', have a great deal to offer to the child in terms of learning problem-solving and logical structures, and there is every reason why they should be included as part of the teaching of basic skills referred to above. (The same is true, it should be added, for the teaching of many applications, such as data-bases and word processing, discussed further below.)

For a fourth area of curriculum development, the emphases emerging in Sweden must be noted. It was significant that the FAST Report referred to skills and qualifications required not only in work but also 'in private life'. In respect of the latter, there are two very broad areas for curriculum development: there is that area of civic education already mentioned — future citizens need to understand the principles behind the State's uses of information handling, though what this means in practice will vary from one political system to another. Secondly, there are a number of everyday life-skills which come into being in the wake of the information technologies. Already the developed countries have seen a rapid increase in the proliferation of credit transfer by means of credit cards: first introduced as extensions to banking services, specialized cards are now available for shops, garages and telephone payments. The development of EFTPOS (electronic funds transfer at point of sale) services will further reduce the use of money as an item of physical exchange. This is but a limited example, but it serves to illustrate the point that learning to manage one's personal income in a world of plastic transfer is considerably more demanding than the regulation of the visible coins and notes in a pocket or wallet, and the school curriculum will have to adapt to the teaching of these and similar life-skills.

PROCESSES OF TEACHING AND LEARNING

Despite the relatively long history of use of information technologies in education in both France and the United Kingdom, it has been only over the space of a very few years indeed (as recently, arguably, as 1983 or 1984) that their full power as extensions to the processes of teaching and learning has been appreciated. Early experiments with information technologies as an educational tool attempted to make the computer (for this one aspect of the new technologies was, and remains for the present, pre-eminent in the classroom) replicate in small part the role of the teacher. That is to say, a great deal of early software was of a rudimentary 'drill and practice' type, interrogating the pupil in a given subject area on purely factual knowledge or on the performance of mathematical tasks, and apportioning praise or blame according to the correctness of the response. It is easy — and, fortunately, no longer necessary — to castigate this generation of crude software, which often owed far more to the enthusiasm of teachers for newly-developed programming skills than it did to any real pedagogic insights. Drill and practice certainly has its educational value, within limits, and it is also true that a more sophisticated approach to the monitoring of pupil responses, ideally linked to a master program dedicated to pupil profiling, has a great deal to offer. But it remains true that many of the early programs produced on micro-computers attacked only a very limited area of the teaching learning process, did so in ways which would have been regarded as educationally primitive had not the allure of the new technology been a distraction, and failed utterly to exploit the true potential of the new technologies.

Nor was this a problem associated only with the introduction of the micro-computer. Earlier innovations, such as PILOT in the United States of America, which used mainframe resources, had been infinitely more advanced in their modelling of the teacher-pupil interaction but had demonstrated (O'Shea and Self, 1983) that such modelling is, at the present stage of our capacities to handle the new technologies, essentially limited to instruction in areas which allow of narrow (in the non-pejorative sense) definitions of 'right' and 'wrong' answers. It is an approach, therefore, suited to training far more than it is to education, though the idea lingers on (Steinberg, 1984).

The transition from such early steps to a deeper understanding emerges in the French literature. In the main text of the Simon Report (Simon, 1980), its author quotes, without explicitly challenging, the traditionalist view of the pre-eminence of the teacher in the classroom and is notably cautious about what he calls 'non-directive' computer-aided learning, pointing to the greater demands that it makes upon the teacher if it is to be used effectively. In an annex to the Report (Simon, 1980: Annex 1, pp. 175ff.), however, a collection of five papers argue the case for 'non-directive' computer-aided learning very cogently indeed. H. Wertz states the point of departure very clearly: 'The use of the computer that we advocate is not so much a means of transmitting additional academic knowledge, but rather a way of analysing intellectual processes'. The computer is, after all, argues Wertz, the first tool to automate intellectual functions. It would be very difficult to overstate the importance for education of this simple truth. The main focus of attention, for Wertz and his colleagues, is work with LOGO, and the sentiments which they express tally closely with those of Papert and his pioneering collaborators at the Massachussetts Institute of Technology, summed up in Papert's well-known dictum concerning the importance of allowing the child to program the computer rather than (as in so many computer-aided learning programs) allowing the computer to program the child (Papert, 1980). It is this concept, that of making it possible for the child to utilize the virtually limitless computational power of the technology, that has emerged in the period since the Simon Report was published as the key to the successful insertion of the information technologies into the processes of teaching and learning.

There is a great deal of development work still to be done before we can claim anything near adequacy in our attempts to make pedagogic use of new technologies, but considerable progress has been made, very much in the spirit of the claims made by Wertz and his colleagues. Work undertaken by the Information Technology Group of the Association for Teacher Education in Europe (ATEE), currently nearing completion, suggests that it is now possible to identify the broad categories of pedagogic use to which the technologies can be put. These uses are seen by the Group as a range extending from, on the one hand, 'tools for thinking with' (in the sense in which Wertz and his colleagues refer to LOGO; more recently PROLOG has commanded a great deal of pedagogic attention for its capacity to allow children to construct and explore logical relationships) and 'tools for organizing knowledge' (where are included data-bases, spreadsheets and word processors) to, on the other hand, 'tutorial software' (i.e. programs which have been designed specifically to impart instruction) and drill and practice programs in the category already mentioned. In between these extremes are to be found, for example, material of the 'electronic blackboard' type, and games and simulations.

The ATEE Group paper, which is based on a very considerable collective experience of its members in the training of teachers to work with the new technologies, comes down heavily in favour of the first categories of software in so far as the full exploitation of the new technologies is concerned. The emphasis on 'tools for organizing knowledge' goes beyond the immediate arguments of the Wertz group, but is entirely consonant with them. The crucial point is that they are, after all, technologies of *information* and, while the acquisition of factual knowledge is by no means the be-all and end-all of education, nonetheless the development of the child's increasing ability to master and make sense of the welter of information presented, in a variety of guises, by daily experience is a key goal in the teaching and learning processes. It is here, as Wertz pointed out in 1980, that the impact of the new technologies on pedagogy is revolutionary: information technology offers an automation of, and consequently an extension of, processes hitherto exclusively the preserve of the human intellect.

This is no theoretical dream. The uses of the new information technologies in the management of industry, commerce and government have made the point already that, with them, information retrieval becomes possible on a scale and at a speed hitherto unthinkable; this fact is fundamental to the processes of wealth generation currently employed in the world (Toffler, 1980; Stonier, 1983; Large, 1984). The development of educational applications, inevitably, lags behind that of management uses, but the pattern is already set. The evidence is to be found, for example, in the uses being made of data-bases in the teaching of history: the capacity of a data-base program, using census data from the nineteenth century, to sort through that data, to identify sets and sub-sets and the relationships between them, transforms the role of pupils from that of recipients of information about History to that of junior researchers actually engaged upon 'doing' History. Similarly, word processing software transforms the teaching of writing skills: pupils now have the capacity to work collaboratively if desired, but above all to see the creation of written work not as a once-and-for-all effort but as a task to be worked upon critically, perhaps over a period of time, with the facility to improve and adjust, to incorporate text from earlier efforts or from other sources, which again would be quite impracticable without the aid of the technology. In ways such as these, as well as through the exploration of concepts and logical constructs such as is offered by LOGO, PROLOG and, imminently, SMALLTALK, that the full power of the new technologies is being unleashed in the classroom.

The software exists, and a great deal of experimentation has already taken place. Very importantly, it is not confined to the secondary level; in the United Kingdom at least, where decentralization favours innovation and experiment, much work has been done already with, for example, data-base and word-processing programs at junior (ages 7-11) and even infant (4-6) levels, with exciting results.

First approaches to learning such as these make *considerably increased demands upon the teacher*. This fact was recognized explicitly in the Simon Report, not only by the author but also by the protagonists of the open-ended approach, notably M. Vivet (Simon, 1980: Annex 1, pp. 205ff.), who analyses in depth the nature and the timing of teacher interventions demanded by, for example, the use of LOGO in the classroom. The experience of the ATEE study group is again supportive of this thinking: it is clear that the demands on the pedagogic skills of the teacher rise in direct proportion to the degree of autonomy that the software allows to the pupil. It is relatively easy to treat all pupils in a class as one, drawing on the fiction that all pupils are at the same developmental point and have the same learning needs. It is quite another matter to cope with twenty-five individuals, each armed with computing power and each engaged on his or her own line of enquiry. Group activities are very important in this respect, and there is a great deal of evidence that the use of the new technologies facilitates collaboration and discussion to a marked degree, but to talk of group work is in no way to lessen the demands made upon the teacher's skills.

The second point is very closely allied to the first. So long as we think in terms of tutorial or drill and practice software, or even games and simulations, it is possible to contain the educational use of the new technologies within the traditional picture of the teacher-pupil relationship. At the heart of that relationship is a transactional process between the intellect of the teacher and that of the pupil (deliberately setting aside, for the present, other elements such as creative and inter-personal skills). This is to say that the teacher, in the traditional relationship, is the mediator of what the pupil learns — the mediator, in the sense of 'vehicle whereby' and also arbiter, in the sense that the teacher actually exercises a considerable control over what a child does or does not learn; in systems where teaching relies heavily on the oral tradition, this control is quasi-absolute. But the introduction of the new technologies disrupts this relationship. If we concentrate for the moment on one aspect only of these technologies, namely the micro-computer (communications will be discussed later), we are enabled to view this disruption more closely, in the following terms. It is valid to say that, in the child's process of acquiring knowledge, both teacher and pupil perform information-processing functions. In the simplest terms, the teacher structures and transmits information which the pupil retrieves, replicating the structure to the degree possible. What the introduction of the micro-computer does is to introduce a third partner into the transaction, that third partner being the processing power of the micro. To say this is not to view the computer in animistic terms, nor to claim that the powers of the machine, *overall*, in any way match those of the human being. But just as pocket calculators can vastly out-perform humans in their one restricted function, so a data-base program, for example, can equally out-perform humans in its own specialist domain. Hence, in respect of a given task, the new technologies may well provide a specialist function which is akin to a segment of the human skills of teacher and pupils but which is in fact greater and more efficient than the humans.

The contrast between the television and the computer, from the viewpoint of the relationship with the teacher, is instructive: the television is not disturbing... it is simply another helper in the classroom. Things are very different with the computer: the logic is altered — it is no longer that of the master, but of another master, and the teacher has to be able to cope with the logic of the computer in order to remain in full charge of the class (Schwartz, 1981: 21). Clearly, there is a need for a considerable re-thinking of the teacher role in the context of what is by now the truism that, whereas the technologies of the Industrial Revolution were an extension to human muscle power, those of the Information Revolution are an extension of our mental capacities.

Once again, the pedagogical implications are great. And this leads to the third point to be made, which is that it is increasingly evident that, so far as education is concerned, the issues raised by the insertion of the new technologies are in no sense *technological*, but *pedagogic*. The ordinary teacher need know relatively little about the technology as such, certainly not in the hardware sense: the differences between the instruction sets of Z80 and 6502 processors are not in themselves of the slightest interest to the teacher, any more than is the difference between a capacitor and a universal parallel interface. On the side of operating systems and software, some knowledge is needed - disk management, for example, or finding a route through hierarchical file structures on a network, though here again the (desirable) trend is towards transparency of use. What *is* required of the teacher is pedagogical understanding, the ability to examine a piece of software, or to be inducted into a language such as PROLOG, and to pose and answer the questions: how can I integrate this teaching resource into my classroom activities? what does it offer my pupils? what questions of classroom management does it invoke? what teaching goals does it offer me that are not attainable (or not as well obtainable) by other means?

The fourth point gives the measure of that challenge, in terms of *processing power*. It is, quite simply, that we are only at the beginning. Development in the United Kingdom has been very largely based around an 8-bit machine which, though versatile and capable of being expanded, remains essentially limited by the low memory capacity which is in fact built in to 8-bit machines. In France, the machines used in schools have tended to be more in the 'business machine' category, but the hardware used in both countries, and in North America as well, is already being overtaken by new generations of machines. The pace of technological advance in this field is breathtaking, and can be gauged from BYTE magazine's chronology of American hardware: in 1975, the Altair 8800 'mini-computer' was launched in kit form with 256 bytes of memory; in 1977 both the Commodore Pet and the first Apple appeared, each with 4K of RAM; IBM entered the micro-computer market in 1981 with the personal computer and its 64K of RAM; while the talked-about machine of early 1986 is the Commodore Amiga which, with a reported theoretical capacity to address 8 Megabytes of memory, has 2,000 times the capacity of the Pet and the Apple (*Byte*, Vol. 10, No. 9, September 1985). This progression of course reflects the chain of development from the early 4-bit to 8-, 16- and currently 32-bit processors. This is a progression which will inevitably continue, with all the increase in portable processing power that such development entails — and already we have first trials of a totally new concept in processing, namely the transputer, which is a complete computer on a single chip, capable of being linked into a very large network wherein each transputer executes its own specified range of tasks in parallel with the processing being done by other transputers in the system (as distinct from current von Neumann architectures in which one central processor carries out all instructions in sequence). The transputer concept is an extremely powerful one, and it illustrates the scale of the problems which will have to be faced by educators attempting to respond to the challenges of the new technologies in the 1990s: the myth of the child carrying to school a micro with a processing capacity equivalent to that of the most powerful main-frame known today will cease, quite simply, to be a myth.

Finally, while it is hardware advance that makes possible the sort of increase referred to above, the increase becomes apparent to the user in terms of the software that can be carried by a system, and it is already clear that software development (to use the term rather loosely) will exert a significant influence over aspects of the teaching and learning processes. We have seen already that the introduction into the classroom of data-base programs adds a new dimension to what can be achieved. Until very recently, such programs have been text-based, but the advances of the early 1980s in WIMP (window-icon-mouse programs) have produced a totally new approach to the process of accessing information. Learning to search through text-based data-bases is one process; learning to use a pointing device such as a 'mouse' to open files on-screen, or to work on more than one document simultaneously, is a different process; what is being offered is a different model whereby the structure of information may be conceptualized. Clearly, the introduction of such new models into the teaching and learning processes has important ramifications, not least for the existing teacher corps, and the issues of teacher training will be discussed later. Other examples can be adduced of software advances which, in their various ways, will impact upon the teaching and learning transaction; these include the development of expert systems and of voice recognition. The fact that many of these technologies, impressive as they already are, are still in their infancy is neither here nor there: the important points are that the ten-year-olds of 1995 have been born already and that the majority of the teachers who will be in post in that year (and for that matter, in 2000 as well) are already in post. They — teachers and pupils — will be the daily users of what we now regard as advanced software.

This evolution does not challenge the need for the teacher role; if such challenges do exist, they come from elsewhere in the spectrum of information technologies, as we shall see. What it does is to place a sharply increased demand upon the teacher's pedagogic skills, and the business of teacher training will need to respond at an appropriate level of sophistication. Moreover, the use of the new technologies in the classroom, the making accessible of processing power to the pupil, enforces a radical shift of emphasis from the process of teaching to the process of learning. And this shift may have ramifications also for the structure of the school, as we shall see.

RESOURCES AVAILABLE TO TEACHING AND LEARNING

The question of resources has already been mentioned in connection with the implications of the new technologies for teaching and learning — in referring, for example, to increases in processing power. Indeed, it is impossible to discuss teaching and learning except in the context of the resources available to these two processes. Thus processing power, generally, and specific categories of software (data-bases, word processing, games, simulations) are all resources available to the classroom.

The new technologies will also have an impact upon another, traditionally very important, category of resources, namely that typified by the library and the library book. Many writers have documented the revolutionary impact of the written word, but because we are so conditioned to the use of the written word, it is difficult to grasp fully just how great a revolution the introduction of writing into mankind's intellectual armoury must have been. And it was, of course, a first step only, albeit a resounding one: the invention of printing and the introduction of steam transport (which made possible widespread dissemination of the printed word) were both dramatic steps forward. That such developments have been every bit as momentous for education is self-evident, but it is worth underlining the reasons why this is so: the key point is that the technologies which make possible the printed word and its distribution are technologies of *information*, focused upon its capture, storage and retrieval. For centuries, in fact, the book has been — and continues to be — the dominant technology available to education; it is quite normal for it to be the sole technology available, and it is a cause of shame that, all too often, the resources of even this basic and simple technology are less than adequately available to schools and to pupils. Contemporary technological advance is at least equal in importance to the invention of printing.

In order to examine the future of educational resources through this perspective, it may be helpful to concentrate attention on three aspects of the new technologies: portable memory, the laser video disk, and communications.

The development of external, and therefore portable, memory has been essential to advances in micro-computer technology. The so-called 'floppy disk', such as that on which this book has been prepared, is a technology image that has recently become widely familiar in developed countries. Such disks currently have the capacity to store, on one side, text files equivalent to approximately 100 typed pages of A4 size; these capacities are increasing by a factor of 10 (Johnson, 1983) thanks to progress in the field of perpendicular magnetic recording, for instance. Thus we can already begin to envisage portable resources encoding, for example, all the key information likely to be required at a given level of operation (final-year undergraduate economics, for example, or an English-French dictionary) and capable of being taken out of the briefcase and plugged into any suitable reading device. The development of such resources will have taken advanced mankind full cycle, in terms of reliance on memorization, from our origins in purely oral traditions. And this is to take only a limited view of the technological possibilities. It is confidently predicted that research will lead to the production of portable memory devices, of shirt-pocket size, able to contain the equivalent of an entire library of books.

While such devices are in a certain sense unthinkable, it is possible to see their implications for the goals of education; the contrast with oral cultures is useful here. In such cultures, the totality of knowledge is constrained by the limits of human memory, both for the individual and for the collective. Thus the development of memory skills attains a very high priority. Indeed, it could reasonably be argued that this priority has engraved itself sufficiently upon the human consciousness for its influence to persist in education to a greater degree than is justified in an age of print technology. What electronic memory capacities of the size just described must inevitably do, however, is to call in question the importance of memorization *at least of factual knowledge*. There is here the interesting paradox, that the more the educational use of the technology is developed, the less important becomes the fact that it is *technology*, and the more important becomes its humanizing impact. Put simply, this is to say that if the technology is capable of coping with the tasks of memorization that we set ourselves, then education can concentrate on the development of the higher-level skills of accessing and of interpreting the information stored; this may well be one of the most important of the long-term consequences of introducing the new technologies into education. What we certainly do face is a major increase in the library and archive resources available to the learner, not only in the formal learning situation but in effect anywhere where it can be 'read'.

A second aspect of external resource development concerns the nature of the information stored. So far, the tendency has been to think of this in text-based or numeric terms, as provided by data-bases and spreadsheets. The development of the laser video disk opens up a new kind of storage, this time of still or moving images, with the advantages over traditional methods of storing such images (photographic collections, slides, films) of large storage capacities — which are constantly being increased — and of rapid access to selected parts of the resource, given the ability to interface disk players with computers and thereby with control programs. In principle, the laser disk opens up a totally new dimension of visual resource, drawing on the wealth of material created since the invention of photography and of cinematography, as well as upon the ability to make a visual record of archive material in other media.

Already the first steps in these directions have been taken (and, as so often, interesting work has been done at primary school level), and there can be little doubt that the laser disk is set to add yet another revolutionary dimension to our understanding of the term 'educational resource'. What is pedagogically interesting relates to what has been said earlier about icon-based interfaces, which offer new conceptual models of the structure of information; here, too, a great deal of thought must be given to ways in which visual information can be structured so as to make it easily accessible. There is also a major reservation to be considered which stems both from the wealth of existing material and from the complexity of the visual information it contains. If we take the seemingly (at first sight) trivial example of a 1930s Hollywood 'B' movie, and consider this simply as a visual resource, then we see that it could be tagged for searching in respect of visual data on topics as diverse as motor cars, dress, architecture, speech idioms, eating habits, portrayal of crime and law enforcement, agriculture, social conditions, biographical material on actors and actresses, directorial style, cinematographic techniques — nor is this list complete. And that is a 'B' movie alone, of which there were thousands, not to mention the historically invaluable documentary material that has been created, and is being added to constantly. The wealth of resource here, not only in our own eyes but from the point of view of future generations, is enormous; the task of transferring it to disk and of providing it with structured tags for access is beyond comprehension, even on a highly selective basis. Yet this is the nature of the resource opportunity made available by just this one aspect of the new technologies.

The educational use of the laser disk, however, has begun already. It is interesting that, as in the case of much software development, there is much to be learned from industrial and commercial use — the disks, for example, are widely used for the training of sales representatives and persons in similar occupations. In this way, the disk becomes a medium for relatively straightforward computer-aided learning usage. But it is the resource use that is potentially the most exciting: already there is at least one encyclopaedia stored on laser disk, while in the United Kingdom the ambitious Project

'Domesday' aims to put onto disk a comprehensive text- and image-based account of contemporary England (to celebrate the 900th anniversary of the inventory carried out after the Norman Conquest). The number of disks available which allow of control via the computer is at present small, but growing, and this aspect of the technology appears set to make a major contribution to educational development in the 1990s.

In addition, the communications aspects of the new technologies open up yet a further dimension of possible resourcing for education. The idea of education at a distance is in no sense new: the 'old' new technologies of radio and television have long been used for this purpose. The limits upon such technologies in the past, however, have stemmed from the relatively few transmission channels available at any one time and from the essentially unidirectional nature of transmission. New communications technologies (including fibre-optic cable, if and when this becomes an economic feasibility, but far more probably satellite transmission technology) introduce, yet again, a new dimension. In the first place, the new systems can carry a considerably increased number of transmission channels open simultaneously; secondly, they allow interactivity (as, for example, in the case of the multi-user adventure games already being played via public telephone systems).

Many different possibilities open up. One relates to what has been said already about portable memory and what it entails in terms of accessing information. Through communications technology, it becomes possible for schools, classes or individual learners to access remote data-banks, of which the American DIALOG is perhaps the best known. This mega-store of electronic information is made up of 200 separate data-bases, with a known total of 90 *million* items stored. Yet, this vast resource can be accessed with a home computer, via the (worldwide) public telephone system; once a search request is properly entered, it is answered within seconds.

The implications for teaching and learning are clear. Such resources act as an inexhaustible memory store. Used properly, they put the educational emphasis entirely upon the skills of accessing and using the information stored. That is to say that, as with other aspects of the introduction of information technologies into educational use, the effect is not to reduce education to a mechanist process, but to further humanize it.

CLASSROOM AND SCHOOL

It is important to stress that nothing that has been said so far concerning technological development rests on flights of fancy. The majority of the technological advances referred to have already taken place, and only in a few cases — for example, when dealing with the capacities of external storage devices — has it been necessary to refer to stages of development which have not yet been achieved. Even in these cases, moreover, the developments mentioned are those known to be on-stream for delivery in the near- to mid-term future. In some areas of potential development, however, the take-up depends, not on technological feasibility, but on a range of other factors, and these area of critical importance. These include:

— the extent of what might be termed the collective pedagogic imagination: it is relatively easy, as we have seen, to subordinate the full potential of the new technologies to a traditionalist view of the teaching learning processes; it will take a very considerable effort of pedagogic imagination and re-thinking to absorb fully the possibilities offered by the new technologies;

— the extent of the political will to affect the changes which are possible and — coupled with this as a necessary consequence — the extent of preparedness to find the economic resources needed for a renewal process that is expensive in terms of capital investment, in curriculum and materials development and, perhaps above all, in the initial and (particularly) the in-service training of teachers;

— the extent and nature of the socioeconomic and cultural change brought about by the insertion of new technology generally in the overall life-style of a country, and the impact of such change on the context within which the education system is placed.

Obviously much of what has already been said has implications for the settings in which, traditionally, education is delivered. These settings follow the pattern set by tribal elders and Socrates equally: that is to say, they are dominated by the need to assemble the learners, in greater or lesser number, within hearing distance of the fount of wisdom, the teacher. This pattern can absorb the need for work to be undertaken away from that fount, temporarily (the Biology student must move away to look, alone, down the microscope, just as the young hunter must eventually take his own spear in hand), but this does not alter the role of the teacher as the point of reference, the arbiter of the learning process. Clearly, this is a crude caricature, but it represents the dominant in the tradition as we know it.

Recent developments, in many cultures, have served to weaken the hold of the teacher-centered approach, and the child, and the learning needs of the child have been much more prominent in modern thinking about education. The question now posed by the information technologies is that of the extent to which we would want to see a further move from the teacher-centered to the pupil-centered approach and, with such a move, the possible erosion of the classroom and the school as we have known them. It is already well established that a great deal of learning is informal (peer learning in the playground and later in the work place, the influence of television in the home), and there have been many experiments with non-formal (i.e. alternative) settings for learning. Now, however, the communications aspects of the information technologies make it possible to envisage a very considerable extension of both informal and non-formal resources, as well as allowing for new modes of delivery to the (more or less) traditional classroom.

The key technological advances are those of portable memory (as we have seen — with implications especially for informal learning) and the twin areas of satellite communication and improved telephone services, the latter including digitization and packet-switched streaming. Finally, the first of a new generation of television receivers, incorporating a degree of intelligence and capable of being tuned to up to a hundred channels, are beginning to emerge.

Against this background it is tempting to toy with thoughts of a vast — and international — store of resources for learning, not merely the informational resource provided by DIALOG and the like, but also of simulations, educational games, sophisticated computer-aided learning modules, shared data-bases and freely down-loadable utilities such as word processing available for several levels of learning ability — all capable of being accessed through the home computer terminal (for in this kind of world the micro/mainframe distinction is meaningless) and displayed on a sophisticated television receiver. The key word here, of course, is 'home', and whether such a picture (which is technologically quite possible) represents a desirable way ahead, with all that it implies for inequality of opportunity between rich and poor even in one cultural setting, is highly debatable, though at least one writer sees it as a likely development route:

'The major shift in technology, then, involved the emergence of the electronic home-based education capability using the television screen and the computer terminal at its base, tied via telephone, airways or local cable television to the local education authorities, a second tie-in to national and international computer network systems and finally a tie-in to the global library archives... traditional subjects can all be learned at home... tailor-made, individually oriented education will be able to replace the much less adaptable mass classroom-based education currently imposed on children in Western countries... During the first decade of a child's life, most of its learning will involve learning in its own or a neighbour's home (Stonier, 1983: 173-4).

This picture of the future may be seen as a relatively radical view of likely development, and it is clear that there are a great many arguments which could be brought to bear against its eventual realization. These arguments include social undesirability (both locally and internationally), the educational undesirability of educating children in isolation from all but a (socially narrow) group of their peers, and finally the need for very heavy investment in resource development before anything resembling a total learning package could be delivered in this way.

What is true for the classroom is true also for the school. And once we accept that the role of the school *vis-à-vis* its 'own' pupils is modified in the direction of greater flexibility of work patterns, heavily based on the provision of access to resources, then it is but a short step to seeing it not as an institution serving one restricted group of pupils, but rather as a base for learning that could and should be open to the whole community to suit the learning needs and schedules of a far wider clientele. Already such loosening of structures is taking place, as exemplified by, for example, the British Open University and, more recently, 'Open-Tech'-type institutions. And there is no reason why the word 'open' should mean 'open within a given national context'. The classic German concept of *Lernfreiheit* — that is, the freedom of the student to follow classes at a number of different universities, over whatever period of time he or she chooses — is one that could well be married with the facilities provided by the information technologies to open up even a transnational approach to learning modules. If, that is, we wish to go in that direction; we have now entered an area dominated by political and economic considerations.

While the scenario presented here may seen in some respects as fanciful, it must be insisted that it represents no more than the technology is capable, already, of delivering. And it is worth pointing to one factor which may be urging development along the lines indicated here: cost-effectiveness. The fact of the matter is that good electronic resource material is expensive to produce (we are, fortunately, a long way away from the early misconception that any teacher with some knowledge of BASIC could produce a worthwhile program in an evening or two). A very high investment of time and of skilled input to design and coding, coupled with rigorous testing over a period of time in schools, is called for before any piece of educational software should be released. The production of video disk material, especially, is likely to remain high — the determining factor, always, being the human rather than the technical costs. Since the educational market, compared to that of business and management, is relatively small in many countries, it is not particularly attractive to market forces and it is heavy in costs to public or national bodies, hence there are attractions of scale, at least, in taking a transnational approach to resource production, not to mention arguments based on the intrinsic merits of education cooperation.

However, there is one, potentially negative, side to this picture of development. It is that, as the preparation of the resource material becomes more expensive and more complex, requiring perhaps long production schedules involving a number of participants, so it becomes more and more removed from the control of the classroom teacher. At first sight this may not seem to matter much, since the classroom teacher is not in fact in control of, say, textbooks. But there is an important difference between the new resources and books, or even educational television. The difference resides in the interactivity of many of the new resources. Both the book and even the television programme are largely passive resources, leaving the teacher at liberty to ignore passages or choose sections out of sequence in the case of the book or to decide what to discuss not discuss (or even to switch off) in the case of television. In the case of information technology-based resources, what is provided is, often, not a passive resource but a learning strategy as well; often this strategy can be highly directive. Thus we have an ambiguous situation. On the one hand, teachers may often and legitimately resent this intrusion into 'their' classrooms; on the other, there is a potential for disseminating advanced teaching strategies which simply would not be possible without the aid of the new technologies. In either case, this involved new definitions of the concepts of classroom and of school.

SOCIOECONOMIC AND CULTURAL CONTEXTS

The uncertainties surrounding the precise effects of the information technologies have been rehearsed in an earlier chapter and need not be repeated here. It is clear that in the developed countries, information handling and processing will be involved in both the generation of wealth and in the servicing of society, with the implication that the teaching of information skills will be one of the fundaments of modern learning. It is difficult to conceive of any context in which this might become a mistaken investment. It is not even a matter of information skills being required only by the most advanced workers, since a great many relatively low-level tasks will demand, increasingly, at least a basic knowledge of information handling. Intriguingly, if there is any cloud on this particular horizon, it is one that may have an adverse effect on some high-level employments: it will be interesting to see, for example, what effect the development of expert systems has on the medical and legal professions. But that, for the moment, is to look further ahead than is reasonable, given our present state of understanding. The implication for education systems in the developed economies is clear: investment in appropriate strategies for the insertion of the information technologies in education is and will continue to be a high priority.

This conclusion holds true, even if the more pessimistic forecasts of the impact of new technologies on employment levels are proved correct, because of the role of information technologies in leisure and in systems of social control. Whatever pattern of employment that eventually emerges, consideration will have to be given to the concept of leisure. In the context of the current situation, with high levels of unemployment especially among the young, the word would be a euphemism, but it represents a concept which may have to be taken more seriously at the cost of rethinking some educational goals. These at present rest on the assumption that pupils will move, on completion of the compulsory cycle of education, either into further education or training or else into tertiary education. While creative and athletic skills are developed at school, they do not fit into any framework of thinking about the use of leisure time. But given the qualitative difference between the concepts of leisure and idleness, it would be necessary to introduce such a framework into the curriculum. It is not difficult to identify the components of such a framework; they include a training in making decisions for a creative use of leisure time, development of the personal skills which would constitute the individual's own resource base and perhaps especially the capacity to engage with the processes of lifelong education which ought to be the corollary of increased leisure. Furthermore, in an ideal society individuals would be able to relate their own skills and leisure time to the wider needs of that society.

Many writers have remarked upon the changes of cultural perception brought about at the level of the individual. Here we may indicate just two possible illustrations of change at this level. The first is already implicit in what has been said about the new pedagogic approaches permitted by the new technologies: what is involved, in effect, is a shift of emphasis towards a very particular set of mental processes associated with the construction of information and the solving of problems. Thus, if the technologies are used properly, information becomes, not a given handed down from the teacher, but an end-product of enquiry, put together from a number of sources, tested, tried out, modified and adapted as the need arises. Such an orientation is nothing new. We might be describing here the process of consulting a dictionary or of using a library, and such processes are of course encouraged in our better schools. The change resides in the extent to which the new technologies facilitate the process of accessing information. They reduce the physical elements of the task (going to the library, handling books as objects, waiting for loan items to be obtained from elsewhere) and emphasize the intellectual process (structuring the enquiry, preparing a set of search criteria, retrieving data, selecting means of presentation). In such processes, the element of construction is important: the structure of knowledge is now more open to control by the user.

In this first illustration, the dimensions of change are many and subtle. The idea of constructing an algorithm is different in kind from that of adopting an ordered approach to solution-finding by manual means. The point about an algorithm is that it presupposes the automation of a significant part of the intellectual process. This means that human ingenuity is directed, not towards finding the solution itself, but towards constructing a set of automation procedures which can be guaranteed to find the solution, no matter how many iterations are called for. We might put this another way by saying that to use the new technologies is to accept the principle of an extension to our intellectual capacities as human beings. This, clearly, is an important modification to our shared cultural perceptions. The development of algorithms, the construction of multiple-key search criteria for data-bases, the use of a high-level language for coding, or of heuristics, and even the playing of electronic games all imply this element of dialogue between the human intellect and the formal structures which are the key to the new technology. Those formal structures, of course, have been created by human intelligence, but they have been created to utilize to their maximum potential the processes of which machines are capable. Thus access to the technology is intermediated by the requirements of the machine, and it seems inevitable that our human habits of thought will modify themselves as we learn more and more to use that access.

The second illustration is closely allied to the first, and points us back from the level of the individual to that of the macro system. As our intellectual habits modify, so also do our perceptions of these habits. To use a word processor, linked to communications facilities, for the production of text is to acquire a new perspective upon one's capacity to write, if not in terms of ultimate quality then at least in terms of control, of structure, of planning and even of practicalities of delivery. The same is true of the use of data-bases for relatively mundane tasks: the effect upon the user is a liberating one. When we consider the domestic or hobbyist use of the information technologies, we encounter a concept of user autonomy akin to the learner autonomy created in the field of education, simply in the sense that the technologies permit the individual a far greater penetration of the total information resource than ever before: the phenomenon of the 'hacker', for example, is one that would repay serious scrutiny as representing a totally new model of relationship between individual and the macro system. Many writers — among them Masuda and Toffler — have seen such developments leading to a new symbiosis of individual and society; if realized, such developments have interesting implications for the governance of society.

It will be clear that the scenarios sketched here are tentative to a degree, and that it is, and will remain, impossible to make precise predictions of cultural change of the kind that would enable us to make tidy planning provision. Moreover, everything said hitherto in this chapter has drawn upon the context of the societies and cultures of the developed countries in which the new technologies have so far made their impact, both generally and specifically in education. To turn to the needs of, and challenges to, the developing world is to raise a range of even wider and more searching questions.

New technologies, education and the developing countries

The overall impact of information technologies upon education in the developed countries, even within Western Europe, reveals wide differences of orientation between countries with broadly similar socioeconomic settings, so that it is impossible to generalize about 'the developing world' and the new technologies: as has been emphasized already, there are a great many developing countries in that world, and their situations, needs, cultures and developmental paths are very different one from the other. Nevertheless an attempt must be made to examine the consequences of the possible take-up of the new information technologies by the education systems of developing countries.

If we look at the impact of the information technologies on, for example, the curriculum and particularly on the new job-related skills that will be needed in developing countries, it is easy to conclude that information handling skills will be widely required and that, in all educational systems, it will be prudent to lay a sound basis of such skills in the basic cycles of education. There must, however, be a realistic assessment of the context in which these skills will eventually be employed. It is fashionable in developed countries to think in terms of the information technologies taking us into a post-industrial society — an 'Information Society', possibly — and there is the seductive view about that, if only the developing countries can properly acquire and harness the power of the new technologies, especially information technologies, then they will succeed in leap-frogging the industrial stage of development and take their place alongside the already developed nations at the post-industrial level.

The fallacy behind such a view, admittedly over-simplified here, is all too obvious. The only nations capable of moving into a *post*-industrial phase are those which have first undergone the *industrial* phase. As J. Rada points out, ' "information" is the consequence of development and not its cause, although the technology can be used for development purposes' (Rada, 1983). Countries which are rapidly absorbing the methods and structures of new information technologies into their fabric are able to do so, and to sustain the heavy investment involved, simply because they have a highly developed economic and technological base as both a launch-pad and a take-up mechanism. Additionally, they have the resources for development in the shape of universities and research institutes able to make continuing high-level contribution to technological advance. Thus, it may be that the new information technologies, far from being the panacea of underdevelopment, could further exacerbate the gulf between North and South.

There is, for example, the argument already used in respect of the developed countries themselves: whatever the paths of development chosen by a given country, it is difficult to conceive that the process of development would not be enhanced by the fostering of information handling skills. There are some instances in which this is patently true – in the development of health care systems, for example, as well as in the build-up of administrative and service infrastructures which must of necessity relate to a global context characterized by the patterns set by developed countries. In other instances, the link will need to be examined carefully and appropriate strategies constructed; an example would be information technology-based resource management applied to a nomadic environment. It will take a very long time to work out what exactly is and what is not 'appropriate'. In the meantime, it is difficult to see how a degree of investment in the introduction of information technologies into education could fail to be of long-term benefit to a developing country. What we must not expect, quite simply, is that the 'job'-related skills required in a post-agrarian society will exactly match those of a post-industrial society. Information skills, in other words, are going to be as much needed in remote rural communities as they are in high-powered international money markets. The work contexts will be very different one from the other: the skills are basically the same.

The potential benefits of the information technologies become even clearer when we consider them as a basis of educational methodology, of resourcing and of ensuring a delivery system. It is difficult to think of any reasons, apart from the potentially very important ones of cultural inappropriateness (a matter which will be addressed later), why everything said about the take-up of the information technologies in the education systems of the developed world should not apply with equal force to those of the developing countries. Furthermore, it can be argued that some of the possibilities identified point to an even more valuable exploitation in the developing world than in the developed.

We saw that the information technologies make possible the (central) creation of sophisticated learning packages and resources. In some countries of the North this might be seen as a challenge to the cherished autonomy of the teacher, but in countries where trained teacher supply is still a problem, then the argument for a lead from the centre becomes very strong. In addition to the possible learning packages themselves, two further areas in which the technology could be supportive are those of remote communication, probably by satellite, and interactivity. The potential is very high on a number of scores. Given what has been said about the increased demands made by the use of new technologies on the pedagogical sophistication of the teacher, an appropriate strategy in a number of cases might be to start with the use of communications capabilities for the distribution of essentially traditional-type instruction, so as to tackle basic needs. But as in the case of the developed countries, the installation of the communications facility within the school makes it an 'open' institution, capable of being used as a community learning centre. In this way, the new technologies become a modality not only for the training of pupils but also of training teachers in-service, community leaders, health workers – given imagination and adequate resourcing, there is no limit to what could be achieved. And what was said about the school as access point to the global information base holds as true for the developing as for the developed world.

Many experiments, of course, are already *in situ*, and experience in India with satellite (INSAT-1B) dissemination of educational programmes is illuminating. D. Swaminathan (1985), among others, is categoric in his assertion that 'mass communication is an essential catalyst for national development and social change', but he does not shrink from identifying the problems. One, a well-known problem from many development education projects, is the need for appropriate training. Merely to make available the technology is not enough, since staff have to be trained in the production of the requisite materials, in both the educational and the technical senses. At present, Swaminathan reports, the lack of indigenous materials and a shortage of trained personnel forces Indian users to rely on materials produced abroad, such as British Open University programmes. His remarks were made in the context of higher education; clearly, the observations are equally applicable to any level of education.

The picture is by no means all negative. The new technologies do offer unparalleled opportunities — via international packet switched-stream (IPSS) telecommunications — for collaboration, and it is important that such opportunities be explored constructively and sensitively. Yet problems such as those of language are persistent, and far-reaching. The indications are there already in the developed world. In France, for example, there early emerged a strong concern that English words provided the basis of high-level programming languages and, although this was not a barrier for the children learning to use the languages in question (principally, BASIC), there was nonetheless official concern for the predominance of English over French, and rightly so. One consequence was the design of a French-based language, LSE, for school use.

The problem is compounded for a country such as India, where it is not so much a question of a language (in this case, English) forcing itself upon a recipient country from the outside — or at least, no more so than has happened historically in the relationship between England and India. In this case, it is the multiplicity of languages used on the sub-continent that has led Indian authorities themselves to decide that English shall remain for the present the *lingua franca* of electronically disseminated education. As R. Amritavalli puts it, 'The real problem is not a choice between English and some other language, for that...is no choice at all'. A major problem is that of linguistic proficiency; as Amritavalli points out, a semi-proficiency that makes it possible for television narrative to be followed adequately, or adequately enough, just will not suffice when the material being broadcast is expository. Until the problem of linguistic proficiency is solved, if ever it is, it is essential that a presentational style be adopted which does facilitate comprehension (Amritavalli, 1985).

This is to touch only the problems of spoken language. Difficulties multiply when we are faced with the demands of the written word which almost invariably must support the spoken, and which become acute when non-Western scripts are used; the difficulties experienced in one of the most technologically developed of all countries, Japan, in producing a computer 'keyboard' capable of handling *kanji* characters, are well known. It is difficult to see how roman script can be prevented from being the *de facto* world standard for computer keyboards, with all that that implies for other scripts and for the centuries — in some cases millennia — of cultural development behind them.

And of course the problem goes deeper. We have seen also how the use of the procedures of information processing, geared as they are to what the machine is capable of, are likely to modify our own processes of thought, our intellectual tools for modelling reality. But the developed world has the advantage that the human input to the development of the information processing has come from within its own cultural traditions; if we face cultural change, then at least we set out from our own base point. This is not the case for the developing countries, for which information processing is, *ab initio*, a given handed down by the developed world. For them, cultural interference is present from the beginning. That the potential interference goes deep is evident: what is involved here are modifications to the structure of thought processes, to language as the expression of those thought processes, to accepted patterns of the representation of knowledge, to the role of accepted knowledge as determinant of the social hierarchy, to the social role of the teacher. G. Shanmughasundaram (1985) reminds us of the centuries-old role of the *guru* in Indian society. It is a role which is fundamental to the very fabric of that society, but it is not a role than can compete with the information-handling capacities which the new technologies make available.

Changes in these and similar areas are of a major order of importance. It is not difficult to generate enthusiasm, within the cultural context of a 'developed' education system, for the potential offered by the new technologies for learner-centered learning and for learner autonomy, with all that that implies for re-assessment of the role of knowledge in the social hierarchy. It is easy to think of religious and social contexts in which such enthusiasm would be entirely wrong. The threat of misplaced initiatives leading to unwanted results is a real one.

Yet the dilemma is equally real: not to make the power of information technology available to all education systems to use as they see fit would be inexcusable. And there is the further, ironic thought, that the intellectual processes of the cultures of the North have no monopoly of wisdom: might there be, in the thought-processes nurtured in other cultures, insights and modes of operating which could make, conceivably, important contributions to our further exploration of the new technologies? What has Artificial Intelligence research, for example, to learn from the myth-retention mechanisms of oral cultures? Certainly it will be interesting to discover whether the work currently underway in Japan in the field of AI will have amongst its outcomes procedures and processes which will prove to be identifiably not American, not European.

There are, inevitably, unanswerable questions. But what matters, urgently, at this stage in our understanding of the new technologies and of what they imply for education, is that we do our utmost to ensure that at least the questions asked are the right ones. The discussion here has focused on what the new technologies appear to signal for education in the developed world, in the hope that the analysis will help to pinpoint some of the problems which they raise specifically for the developing countries. What emerges, above all, is the question of cultural change, since that lies at the heart of the impact of the new technologies.

References

AMRITAVALLI, R. 1985. The Language of Television, Visualising the Verbal By-Pass. In: International Conference on New Technologies in Higher Education, 28-9 November 1985. New Delhi.

CERYCH, L. 1985. Problems Arising from the Use of New Technologies in Education. *European Journal of Education*, Vol. 20, pp. 223-32.

FAST. 1984. *Eurofutures: The Challenges of Innovation.* The Commission of the European Communities.The FAST Report. Borough Green, Sevenoaks, Kent (England), Butterworths.

FOTHERGILL, R. 1981. *Microelectronics Education Programme: the Strategy.* London, Department of Education and Science.

GWYN, R. 1984. *New Teaching Functions and Implications for New Teaching Programmes.* Paris, Organisation for Economic Cooperation and Development.

JOHANSSON, S.A.; KOLLERBAUR, A.; BOLLANDER, L.; DAHLBOM, B. 1985. *Policy and Points of Departure — Education for the Computer Age.* Stockholm, Swedish National Board of Education.

LARGE, P. 1984. *The Micro Revolution Revisited.* London, Frances Pinter.

MASUDA, Y. 1983. *The Information Society as Post-Industrial Society.* Bethesda, MD, World Future Society.

OECD, 1985. *New Information Technology: A Challenge to Education.* Paris, Organisation for Economic Cooperation and Development.

O'SHEA, T.; SELF, J. 1983. *Learning and Teaching with Computers.* Brighton, Harvester Press.

PAPERT, S. 1980. *Mindstorms: Children, Computers and Powerful Ideas.* Brighton, Harvester Press.

RADA, J. 1983. *Information Technology and the Third World.* Vienna, IFAC Seminar.

SCHWARTZ, B. 1981. *L'informatique et l'éducation.* Paris, La Documentation Française.

SHANMUGHASUNDARAM, G. 1985. Education as a Technique of Transmitting Civilisation: Impact of New Technologies. In: International Conference on New Technologies in Higher Education, 28-9 November 1985. New Delhi.

SIMON, J.C. 1980. *L'éducation et l'informatisation de la société.* Paris, La Documentation Française. Annexe 1: *Les voies de développement. Contribution des Groupes de Travail.*

STEINBERG, E.R. 1984. *Teaching Computers to Teach*. Hillsdale, N.J., Lawrence Erlbaum Associates.
STONIER, T. 1983. *The Wealth of Information*. London, Methuen.
SWAMINATHAN, D. 1985. Utilisation of INSAT Facility for Higher Education in India. In: International Conference on New Technologies in Higher Education, 28-9 November 1985. New Delhi.
TOFFLER, A. 1980. *The Third Wave*. New York, Morrow.

Selective bibliography

I. General references

ARAD, R.W.; MCCULLOCH, R.; PINERA, J.; HOLLICK, A.L. (eds.). 1979. *Sharing Global Resources*. 1980's Project, Council on Foreign Relations. New York, McGraw Hill.

BRESSAND, A. (ed.). 1981; 1982; 1983-84. *Ramsès*. Rapport annuel de l'Institut français des relations internationales, IFRI. Paris, Economica.

BROWN, L. (ed.). 1984. *State of the World 1984*. Worldwatch Institute. New York, Norton.

BUHR, M.; KRÖBER, G. (eds.). 1977. *Mensch — Wissenschaft — Technik*. Berlin, Akademie-Verlag.

COLE, S.; MILES, I. 1984. *World Apart : Technology and North-South Relations in the Global Economy*. Brighton, Sussex, Wheatsheaf.

COLOMBO, U.; TURANI, G. 1982. *Il secondo pianeta*. Milan, Mondadori.

DUCROCQ, A. 1984. *Le futur aujourd'hui. 1985-2000. Les quinze années qui vont changer votre vie quotidienne*. Paris, Plon.

Economie mondiale. La montée des tensions. 1983. Paris, CEPI, Economica.

FORTI, A. (ed.). 1984. *Scientific Forecasting and Human Needs. Trends, Methods and Message*. Proceedings of a symposium held in Tbilissi, USSR, 6-11 December 1981. Paris, Unesco and Oxford, Pergamon Press.

FREEMAN, C. (ed.). 1983. *Long Waves in the World Economy*. Sevenoaks (UK), Butterworth.

GAPPERT, G.; KNIGHT, R.V. (eds.). 1982. *Cities in the 21st Century*, Vol. 23, Urban Affairs Annual Review. London, Beverley Hills, New Delhi, Sage.

The Global 2000 Report. Washington, D.C., Council on Environmental Quality, U.S. Department of State, Vol. 1, 1980 : *Summary Report*; Vol. 2, 1980 : *Technical Report*; Vol. 3, 1981 : *Documentation on the Government's Global Sectoral Models. The Government's "Global Model"*.

Global Future : Time to Act. 1981. A report to the President on global resources, environment and population council on environmental quality. Washington, D.C., U.S. Department of State.

Global Models, World Futures, and Public Policy : A Critique. 1982. Washington, D.C., Office of Technology Assessment (OTA), OTA-R-165.

GOLDSMITH, M.; KING, A. (eds.). 1979. *Science, Technology and Global Problems. Issues of Development : Towards a New Role for Science and Technology*. Oxford, New York, Pergamon Press.

GVISHIANI, J. (ed.). 1979. *Science, Technology and Global Problems*. Proceedings of the international symposium on Trends and Perspectives in Development of Science and Technology and their Impact on the Solution of Contemporary Global Problems, held in Tallinn, USSR, 8-12 January 1979). Oxford, New York, Pergamon Press.

HARTMANN, K.; FISCHER, I. 1980. *Technologie, Wachstum, Produktivität*. Berlin, Dietz Verlag.

HAQ, K. (ed.). 1983. *Global Development : Issues and Choices, North/South Roundtable*. Washington, D.C.

KAHN, H. ; SIMON, J. 1984. *The Resourceful Earth : A Response to Global 2000*. New York, Blackwell.

KOTHARI, R. *Footsteps into the Future. Diagnosis of the Present World and a Design for an Alternative*. 1974. New York, The Free Press.

–. (ed.). 19... *State and Nation Building. A Third World Perspective*. New Delhi, Allied Publishers.

LEONTIEF, W. *et al*. 1977. *The Future of the World Economy*, Oxford University Press.

–. 1977. *1999 : l'expertise de Wassily Leontief*. Paris, Dunod.

–. *Le Pacifique, "Nouveau Centre du Monde"*. 1983. Paris, Berger-Levrault.

LESOURNE, J. 1976. *Les systèmes du destin*. Paris, Dalloz.

–. 1981. *Les mille sentiers de l'avenir*. Paris, Seghers.

LESOURNE, J. *et al*. 1979. *Facing the Futures : Mastering the Probable and Managing the Unpredictable*. Paris, OECD.

LONG, F.A.; OLESON, A. (eds.). 1980. *Appropriate Technology and Social Values : A Critical Appraisal*. Cambridge, Mass., Ballinger Publishers.

Man, Science, Technology. A Marxist Analysis of the Scientific and Technological Revolution. 1973. Moscow, Prague, Academia Prague.

MATTIS, A. (ed.). 1983. *A Society for International Development: Prospectus*. Durham, Duke University Press, Duke Press Policy Studies.

–. *Nord/Sud. Du défi au dialogue*. 1978. Paris, Shed-Dunod.

PAPON, P. 1983. *Pour une prospective de la science. Recherche et technologie : les enjeux de l'avenir*. Paris, Seghers.

RADHAKRISHNA, S. (ed.). 1979. *Science, Technology and Global Problems. Views from the Developing World*. Oxford, New York, Pergamon Press.

RICHARDSON, J. (ed.). 1984. *Models of Reality : Shaping Thought and Action*. Lomond (Mt. Airy), in co-operation with Unesco (Paris).

ROSENBERG, N. 1976. *Perspectives on Technology*. New York, Cambridge University Press.

–. *Inside the Black Box : Technology and Economics*. 1982. Cambridge, Cambridge University Press.

Science and Technology Indicators. No. 1: *Resources devoted to R&D*. 1984. Paris, OECD. No. 2: R&D Invention and Competitiveness, 1986. OECD.

Science and Technology Policy for the 1980s. 1981. Paris, OECD.

Science and Technology. A Five-year Outlook. 1979. The National Research Council, National Academy of Science. New York, Freeman.

SIMON, J.L.; KAHN, H. 1984. *The Resourceful Earth. A Response to Global 2000.* Oxford, Blackwell.

SYDOW, W. (ed.). 1983. *In die Zukunft gedacht : Wissenschaftler aus 6 Ländern entwickeln Ideen zu Wissenschaft und Technik.* Berlin, Verlag Die Wirtschaft.

The Five-year Outlook on Science and Technology. 1982. Washington, D.C., National Science Foundation. *Source Materials*, Vol. 1 : *Outlook for Science and Technology : The Next Five Years* ; Vol. 2 : *Policy Outlook: Science, Technology and the Issues of the Eighties.*

Tendances à long terme du développement économique. 1983. Paris, Economica, United Nations.

TOFFLER, A. 1980. *The Third Wave.* New York, Morrow.

URQUIDI, V.L. (ed.). 1979. *Science, Technology and Global Problems. Science and Technology in Development Planning.* Oxford, New York, Pergamon Press.

World Population and Fertility Planning Technologies : The Next Twenty Years. 1982. Washington, D.C., Office of Technology Assessment (OTA), OTA-HR-157.

II. Club of Rome publications and related studies

COLE, H.S.D.; FREEMAN, C.; JAHODA, M.; PAVITT, K.L.R. 1973. *Models of Doom : A Critique of the Limits to Growth.* New York, Universe Books. Also published by University of Sussex (Science Policy Research Unit), England, as *Thinking about the Future.*

FREEMAN, C.; JAHODA, M. 1978. *World Futures. The Great Debate.* London, Martin Robertson.

FRIEDRICHS, G.; SCHAFF, A. (eds.). 1982. *Microelectronics and Society. For Better or for Worse.* A Report to the Club of Rome. Oxford, Pergamon Press.

HAWRYLYSHYN, B. 1983. *Les itinéraires du futur. Vers des sociétés plus efficaces.* Paris, Presses Universitaires de France.

KING, A. 1980. *The State of the Planet.* London, Pergamon Press (IFIAS). LENOIR, R. 1984. *Le Tiers Monde peut se nourrir.* Paris, Fayard.

MEADOWS, D.H.; MEADOWS, D.L. ; RANDERS, J. ; BEHRENS III, W.W. 1972. *The Limits to Growth.* Cambridge, Massachusetts Institute of Technology Press.

MESAVORIC, M.; PESTEL, E. 1974. *Mankind at the Turning Point. The Second Report to the Club of Rome.* New York, E.P. Dutton.

OLSON, M.; LANDSBERG, H.H. 1973. *The No-growth Society.* New York, Norton.

III. Prospective studies in science and technology

ADKINS, B. M. (ed.). 1984. Man and Technology. *The Social and Cultural Challenge of Modern Technology.* Newmarket, U.K., Cambridge Information and Research Services.

ALTSHULER, A.; ROOS, D. 1984. *The Future of the Automobile*. Cambridge, Mass., Massachusetts Institute of Technology Press.
BALCET, G.; COLOMBO, U. 1980. *La speranza tecnologica. Tecnologie e modelli di sviluppo per una società a misura d'uomo*. Milan, Etas Libri.
BARRON, I.; CURNOW, R. 1979. *The Future with Microelectronics*. London, Frances Pinter.
BULL, A.T.; HOLT, G.; LILLY, M.D. 1982. *Biotechnology. International Trends and Perspectives*. Paris, OECD.
CHARLES, G.; DYAN, B. (eds.). 1984. *Guide des technologies de l'information*. Paris, *Autrement* ("Sciences et Cie").
CONQUY BEER-GABEL, J. 1984. *Informatisation du Tiers Monde et coopération internationale*. Paris, La Documentation française.
Construire l'avenir. Livre blanc sur la recherche. 1980. Paris, La Documentation française.
GAUDIN, T. *et al.* (eds.). 1983. *Rapport sur l'état de la technique. La révolution de l'intelligence*. Paris, Centre de prospective et d'évaluation du Ministère de l'industrie et de la recherche ; Société des ingénieurs et scientifiques de France, Special number of *Sciences et Techniques* (Nos. 97-8; October 1983).
GOLDSMIDT, P.G. 1982. *Health 2000*. Baltimore, Policy Research Institute, The Health Futures Project.
GOLDSMITH, M. (ed.). 1983. *The Promise of Science and the Limitations of Technology : Interdependence and Self-reliance and the Role of NGO's for Development*. Science and Technology for Development Series, Vol. 3. United Nations ; Dublin, Tycooly International Publishing Ltd.
GLOWINSKI, A. 1980. *Télécommunications : Objectif 2000*. Paris, Dunod. GREENBERGER, M.; CRENSON, M.A.; CRISSEY, B.L. 1976. *Models in the Policy Process : Public Decision Making in the Computer Era*. New York, Russel Sage Foundation.
GROS, F.; JACOB, F.; ROYER, P. 1979. *Sciences de la vie et société*. Paris, La Documentation française.
HANKE, P. (ed.). 1984. *Entwicklungsprobleme und Effektivitätsermittlung in der Biotechnologie*. Berlin, Akademie der Wissenschenschaften der DDR; Institut für Theorie, Geschichte und Organisation der Wissenschaft.
HAUSTEIN, H.-D.; MAIER, H. (eds.). 1984. *Flexible Automatisierung — Entwicklungstendenzen, Probleme, Perspektiven*. Berlin, Akademie der Wissenschaften der DDR; Institut für Theorie, Geschichte und Organisation der Wissenschaft.
HEMILY, P.W.; Ö ZDAS, M.N. 1976. *Science and Future Choice* (NATO), Vol. 1 : *Building on Scientific Achievement*; Vol. 2 : *Technological Challenges for Social Change*. Oxford, Clarendon Press.
HÜTTER, M. *et al.* 1984. *Mikroelektronik und Gesellschaft*. Berlin, Akademie-Verlag.
Industrial Robots : A Summary and Forecast. 2nd edn. 1983. Tech Tran Corporation (134 North Washington Street, Naperville, Illinois 60540, USA).
JOHNSON, D. G. (ed.). 1980. *The Politics of Food*. Chicago Council on Foreign Relations.
Les enjeux technologiques des années 1985-1990. June 1983. Cahiers d'études et de recherche, Commissariat général au plan. Paris, La Documentation française, No. 1.
Les mutations technologiques. 1981. Paris, Association pour le développement des études

sur la firme et l'industrie, ADEFI (VIèmes Rencontres nationales de Chantilly, September 1980), Economica.

Les télécommunications. Perspectives d'évolution et stratégies des pouvoirs publics. 1983. Paris, OECD.

LUCAS, B.; FREEDMAN, S. (eds.). 1983. *Technology Choice and Change in Developing Countries : Internal and External Constraints*. Science and Technology for Development Series, Vol. 1. United Nations ; Dublin, Tycooly International Publishing Ltd.

LUNDSTEDT, S.B.; COLGLAZIER, E.W. Jr. (eds.). 1982. *Managing Innovation. The Social Dimensions of Creativity, Invention and Technology*. New York, London, Pergamon Press.

MOORE, F.T. 1984. *Technological Change and Industrial Development : Issues and Opportunities*. Washington, D.C., The World Bank (World Bank Staff Working Paper, Industry Department).

NATIONAL RESEARCH COUNCIL. 1984. *Genetic Engineering of Plants. Agricultural Research Opportunities and Policy Concerns*. Board on Agriculture. Washington, D.C., National Academy Press.

NORA, S.; MINC, A. 1978. *L'informatisation de la société*. Paris, La Documentation française.

Annexes : I. Nouvelle informatique, nouvelle croissance ; II. *Industrie et services informatiques ;* III. *La nouvelle informatique et ses utilisateurs ;* IV. *Documents contributifs*.

–. 1980. *Computerization of Society*. Cambridge, Massachusetts Institute of Technology Press.

PLESCHAK, F.; KREJCIK, P. 1982. *Automatisierung aus ökonomischer Sicht*. Berlin, Verlag Die Wirtschaft.

RAMESH, J.; WEISS, Charles Jr. (eds.). 1979. *Mobilizing Technology for World Development*. International Institute for Environment and Development, and Overseas Development Council. New York, Praeger.

Rapport sur l'état de la technique. 1983. Special number of *Sciences et Techniques* (CPE/MIR).

SASSON, A. 1983. *Les biotechnologies. Défis et promesses*. Paris, Unesco, Collection Sextant Vol. 2.

–. 1983. *Las biotecnologias : desafios y promesas*. Paris, Unesco, Coleccion Sextante vol. 2.

–. 1984. *Biotechnologies : Challenges and Promises*. Paris, Unesco, Sextant Series Vol. 2.

SIMON, J. C. *L'éducation et l'informatisation de la société*. 1981. Paris, La Documentation française.

Annexes : I. 9Les voies de développement. Contribution des groupes de travail ; II. *Les expériences par pays*.

SMITH, A. 1980. *Goodbye Gutenberg : The Newspaper Revolution of the 1980's*. New York, Oxford University Press.

VOLMER, J. (ed.). 1981. *Industrieroboter*. Berlin, Verlag Die Technik.

WEIZSACKER, E.U. von; SWAMINATHAN, M.S.; LEMMA, A. (eds.). 1983. *New Frontiers in Technology Applications. Integration of Emerging and Traditional Technologies*. Science and Technology for Development Series, Vol. 2. United Nations ; Dublin, Tycooly International Publishing Ltd.

IV. Regional studies

ARAB STATES

Abdalla, I.S. *et al.* 1983. *Images of the Arab Future*. London, United Nations University, Frances Pinter.

ASIA

Rahman, A. *et al.* 1984. *Science and Technology in India*. New Delhi, National Institute of Science, Technology and Development Studies, INSDOC.
Wang Huijiong and LI Po-xi. 1986. *China in the Year 2000*. Research Centre for Economic, Technological and Social Development of the State Commission for Science and Technology, Beijing. (The "Blue Book", with several annexes).

AUSTRALIA

Technological Change in Australia. Report of the Committee of Inquiry into Technological Change in Australia. 1980. Vol. 1: *Technological Change and its Consequences*. Vol. 2: *Technological Change in Industry*. Vol. 3: *Review of Policies and Programme for Technological Change*. Vol. 4: *Selected Papers on Technological Change*. Canberra, Australian Government Publishing Service.

EUROPE

Albert, M. *Un pari pour l'Europe. Vers le redressement de l'économie européenne dans les années 80*. Paris, Seuil.
A Policy-oriented Survey of the Future. Towards a Broader Perspective. 1983. The Hague, Netherlands Scientific Council for Government Policy.
Bona, E. *et al.* 1983. *Future Research in Hungary*. Budapest, Akadémiai Kiado.
Colombo, U.; Lanzavecchia, G.; Mazzonis, D. 1984. *Innovazioni tecnologiche e struttura produttiva - La posizione della' Italia*. Milan, Il Mulino.
Danzin, A. 1978. *Science and the Second Renaissance of Europe*. London, Pergamon Press.
Danzin, A. *et al.* 1980. *La société française et la technologie*. Paris, La Documentation française.
Decoufle, A.C. 1980. *La France de l'an 2000. Une esquisse*. Paris, Seghers.
Eurofutures : The Challenges of Innovation. 1984. The FAST Report, The Commission of the European Communities in association with the journal *Futures*. London, Butterworths.
Godet, M.; Ruyssen, O. 1980. *L'Europe en mutation*. Brussels, Collection Perspectives européennes.
L'industrialisation du bassin méditerranéen. 1981. Actes du Colloque GRESMO, Presses Universitaires de Grenoble.
Nick, H. 1979. *Zur materiell-technischen Basis in der DDR*. Berlin, Dietz Verlag.

LATIN AMERICA

Catastrophe or New Society? A Latin American Model. 1976. Ottawa, International Development Research Centre (IDRC).
SAGASTI, F.R. *et al.* 1983. *Un decenio de transicion. Ciencia y tecnologia en América Latina y el Caribe durante los setenta.* Lima, GRADE.

UNITED STATES OF AMERICA

Information Technology and its Impact on American Education. 1982. Washington, D.C., Office of Technology Assessment (OTA), OTA-CIT-187.

V. Energy

Alternative Energy Futures. An Assessment of U.S. Options to 2025. 1979. Stanford, Institute for Energy Studies.
BUIGUES, P.A. 1981. *Scénarios pour le solaire. Horizon 2000.* Aix-en-Provence, EDISUD.
CURRAN, D. W. 1981. *La nouvelle donne énergétique.* Paris, Masson, Collection Géographie.
DARMSTADTER, J.; LANDSBERG, H.; MORTON, H. 1983. *Energy Today and Tomorrow: Living with Uncertainty.* Englewood Cliffs, N.J., Prentice Hall.
DUBART, J.C. 1981. *Energies : le grand tournant.* Paris, Editions sociales.
Energy in Transition, 1985-2010. 1980. Washington, D.C. National Academy of Science.
FRISCH, J.R. 1981. *Third World Energy Horizons 2000-2020.* A regional approach to consumption and sources of supply. Paris, Editions Techniques et Economiques (published for the World Energy Conference).
–. 1983. *Energy 2000-2020 : World Prospects and Regional Stresses.* London, Graham and Trotman (published for the World Energy Conference).
HILDEBRAND, H.-J. 1975. *Wirtschaftliche Energieversorgung.* Leipzig, Deutscher Verlag für Grundstoffindustrie.
LUNEAU, M. 1982. *Les énergies nouvelles : qu'en espérer ?* Paris, La Documentation française.
PRYOR, A. (ed.). 1983. *Non-technical Obstacles to the Diffusion of Energy Technologies : Proceedings of the Bellagio Conference.* Science and Technology for Development Series, Vol. 4. United Nations; Dublin, Tycooly International Publishing Ltd.
RIESNER, W.; SIEBER, W. 1978. *Wirtschaftliche Energieanwendung.* Leipzig, Deutscher Verlag für Grundstoffindustrie.
WILSON, C. *et al.* 1977. *Energy Global Prospects 1985-2000.* New York, McGraw Hill.
World Energy Outlook. 1982. Paris, International Energy Agency (OECD).

VI. Employment and technological change

ANGELOPOULOS, A. 1983. *Global Plan for Employment and New Marshall Plan.* New York, Praeger.

BLUMENTHAL, M.S. (ed.). 1984. *Computerized Manufacturing Automation: Employment, Education and the Work Place.* Washington, D.C., Office of Technology Assessment (OTA), OTA-CIT-236.

FREEMAN, C. *et al.* 1982. *Unemployment and Technical Innovation.* London, Frances Pinter.

JONES, B. 1982. *Sleepers Wake! Technology and the Future of Work.* Melbourne, Oxford University Press.

LEONTIEF, W.; DUCHIN, F. 1984. *The Impacts of Automation on Employment, 1963-2000.* New York, New York University, Institute for Economic Analysis.

LE QUEMENT, Z. 1983. *L'usine du futur proche. Stratégies internationales d'automatisation.* Paris, Agence de l'informatique (diffusion Hermès).

Microelectronics, Productivity and Employment. 1981. Paris, OECD, ICCP, No. 5.

Microelectronics, Robotics and Jobs. 1982. Paris, OECD, ICCP, No. 7, MILLER, R.J. (ed.). 1983. *New Factories, New Workers.* Special issue of the *Annals of the American Academy of Political and Social Science* (Philadelphia, Pa.). Beverly Hills, Ca., Sage Publications.

NEWLAND, K. 1982. *Productivity : The New Economic Context.* Washington, D.C., Worldwatch Institute, Paper No.49.

NORMAN, C. 1980. *Microelectronics at Work : Productivity and Jobs in the World Economy.* Washington, D.C., Worldwatch Institute, Paper No.39.

ROTHWELL, R.; ZEGFELD, W. 1979. *Technical Change and Employment.* London, Frances Pinter.

SABEL, Charles F. 1982. *Work and Politics : The Division of Labor in Industry.* Cambridge Studies in Modern Political Economics. New York, Cambridge University Press.

Technical Change and Economic Policy. 1980. *Annexes :* I. *The Pharmaceuticals Industry*; II. *The Machine-tool Industry*; III. *The Electronic Industry*; IV. *The Fertilizers and Pesticides Industry.* Paris, OECD.

YANKELOVICH, D.; ZETTERBERG, H.; STRUMPEL, B.; SHANKS, M. 1983. *Work and Human Values : An International Report on Jobs in the 1980's and 1990's.* New York, Aspen Institute for Humanistic Studies.

VII. Methodology

DECOUFLE, A.C. 1978. *Traité élémentaire de prévision et de prospective.* Paris, Presses Universitaires de France.

HETMAN, F. 1979. *The Language of Forecasting.* Paris, Futuribles. *The Study of the Future. An Agenda for Research.* 1977. Washington, D.C., National Science Foundation.

Prospective, prévision, planification stratégique. Théorie, méthodes et applications. 1983. Paris, *Futuribles,* December 1983, No.72.